AF560485

New Generation Fibres

New Generation Fibres

Ajay Shrivastava

RANDOM PUBLICATIONS
NEW DELHI (INDIA)

New Generation Fibres

ISBN 978-93-5111-645-5

Published in 2015 in India by

RANDOM PUBLICATIONS

4376-A/4B, Gali Murari Lal, Ansari Road
New Delhi-110 002
Phone : +9111-43580356, 011-23289044, 011-43142548
e-mail: sales@randompublications.com,
info@randompublications.com, randomexports@gmail.com

Reprinted 2025

Type Setting by : Friends Media, Delhi-110089
Digitally Printed at : Replika Press Pvt. Ltd.

Preface

Fiber is a rope or string used as a component of composite materials, or, when matted into sheets, used to make products such as paper, papyrus, or felt. Fibers are often used in the manufacture of other materials. The strongest engineering materials often incorporate fibers, for example carbon fiber and Ultra-high-molecular-weight polyethylene. Throughout the past 30 years, the communication and information industry has witnessed a progression, some might say evolution, of methodologies employed to terminate optical fibers. As today's enterprise networks and data centers confront bandwidth capacity challenges, high ? density requirements, and ever ? increasing data rates, choosing the correct connectivity method has never been more important to meet the demands for optimum reliability, scalability, performance, time to restoration, and cost effectiveness. As the fiber optic network has evolved, so has the method for connecting fibers. Each method that will be analyzed was conceived in order to improve network productivity, maintain optical path integrity, and maximize cost effectiveness. Unfortunately, as new connectivity methods were introduced, it became clear that the "one size fits all" solution was not the optimum for every network application. Other considerations must be factored into any new connectorization technology while meeting industry standards. First, it must be backward compatible with the installed base; it should be easy to use; and require minimal training or retraining of the installer. Finally, it should be cost effective for the end user.

I would like to thank my team for standing beside me throughout my career and writing this book. My special thanks go to "Random Publications" who have published the book.

–Ajay Shrivastava

Contents

1

Textile and Apparel Industry: An Introduction

The textile industry (known colloquially in the United Kingdom and Australia as the rag trade) is a term used for industries primarily concerned with the design or manufacture of clothing as well as the distribution and use of textiles.

TEXTILE MANUFACTURING BY PRE-INDUSTRIAL METHODS

Textile manufacturing is one of the oldest human activities. The oldest known textiles date back to about 5000 B.C. In order to make textiles, the first requirement is a source of fibre from which a yarn can be made, primarily by spinning. The yarn is processed by knitting or weaving to create cloth. The machine used for weaving is the loom. Cloth is finished by what are described as wet processes to become fabric. The fabric may be dyed, printed or decorated by embroidering with coloured yarns.

The three main types of fibres are natural vegetable fibres (such as cotton, linen, jute and hemp), man-made fibres (made by industrial processes) and protein based fibres (such as wool, silk).

Almost all commercial textiles are produced by industrial methods. Textiles are still produced by pre-industrial processes in village communities in Asia, Africa and South America, as an artisan craft and a hobby in Europe and North America.

YARN FORMATION

WOOL

Wool is a protein based fibre, being the coat of a sheep. The wool is removed by shearing.

Wool is the textile fibre obtained from sheep and certain other animals, including cashmere from goats, mohair from goats, qiviut from muskoxen, vicuna, alpaca, and camel from animals in the camel family, and angora from rabbits.

Wool has several qualities that distinguish it from hair or fur: it is crimped, it is elastic, and it grows in staples (clusters).

Characteristics

Wool's scaling and crimp make it easier to spin the fleece by helping the individual fibres attach to each other, so that they stay together. Because of the crimp, wool fabrics have a greater bulk than other textiles, and retain air, which causes the product to retain heat. Insulation also works both ways; Bedouins and Tuaregs use wool clothes to keep the heat out.

The amount of crimp corresponds to the fineness of the wool fibres. A fine wool like Merino may have up to 100 crimps per inch, while the coarser wools like karakul may have as few as 1 to 2. Hair, by contrast, has little if any scale and no crimp, and little ability to bind into yarn. On sheep, the hair part of the fleece is called kemp. The relative amounts of kemp to wool vary from breed to breed, and make some fleeces more desirable for spinning, felting, or carding into batts for quilts or other insulating products.

Wool fibres are hygroscopic, meaning they readily absorb moisture. Wool fibres are hollow. Wool can absorb moisture almost one-third of its own weight. Wool absorbs sound like many other fabrics. Wool is generally a creamy white colour, although some breeds of sheep produce natural colours such as black, brown, silver, and random mixes.

Wool ignites at a higher temperature than cotton and some synthetic fibres. It has lower rate of flame spread, low heat release, low heat of combustion, and does not melt or drip; it forms a char which is insulating and self-extinguishing, and contributes less to toxic gases and smoke than other flooring products, when used in carpets. Wool carpets are specified for high safety environments, such as trains and aircraft. Wool is usually specified for garments for fire-fighters, soldiers, and others in occupations where they are exposed to the likelihood of fire.

Wool is resistant to static electricity, as the moisture retained within the fabric conducts electricity. This is why wool garments are much less likely to spark or cling to the body. The use of wool car seat covers or carpets reduces the risk of a shock when a person touches a grounded object. Wool is considered by the medical profession to be hypoallergenic.

SHEEP SHEARING

Sheep shearing, shearing or clipping is the process by which the woolen fleece of a sheep is cut off. The person who removes the sheep's wool is called a *shearer*. Typically each adult sheep is shorn once each year (a sheep may be said to have been "shorn" or "sheared", depending upon dialect). The annual shearing most often occurs in a shearing shed, a facility especially designed to process often hundreds and sometimes more than 3,000 sheep per day.

History

Shearers wear moccasins to protect their feet, grip wooden floors well, and absorb sweat.

Up until the 1870s squatters washed their sheep in nearby creeks prior to shearing. Later some expensive hot water installations were constructed on some of the larger stations for the washing. Sheep washing in Australia was influenced by the Saxony sheep breeders in Germany who washed their sheep and by the Spanish practice of washing the wool after shearing.

There were three main reasons for the custom in Australia:

1. The English manufacturers demanded that Australian woolgrowers provide their fleeces free from vegetable matter, burrs, soil, etc.
2. The dirty fleeces were hard to shear and demanded that the metal blade shears be sharpened more often.
3. Wool in Australia was carted by bullock team or horse teams and charged by weight. Washed wool was lighter and did not cost as much to transport.

The practice of washing the wool rather than the sheep evolved from the fact that hotter water could be used to wash the wool, than that used to wash the sheep. When the practice of selling wool in the grease occurred in the 1890s, wool washing became obsolete.

Australia and New Zealand had to discard the old methods of wool harvesting and evolve more efficient systems to cope with the huge numbers of sheep involved. After 1888 machine shearing was introduced, reducing second cuts and shearing time. By 1915 most large sheep station sheds in Australia had installed machines, driven by steam or later by internal combustion engines.

Shearing tables were invented in the 1950s and have not proved popular, although some are still used for crutching.

MODERN SHEARING

Today, large flocks of sheep are shorn by professional shearing teams working eight hour days, most often in spring, by machine shearing. These contractor teams will consist of shearers, shed hands and a cook (in the more isolated areas). The shed staff working hours and wages are regulated by industry awards.

A working day starts at 7:30 AM and the day is divided into four "runs" of two hours each. "Smoko" breaks of a half hour each are at 9:30 AM and again at 3:00 PM. The lunch break is taken at 12:00 PM for one hour. Most shearers are paid on a piece rate, *i.e.*, per sheep. Shearers who "tally" more than 200 sheep per day are known as "gun shearers".

Typical mass shearing of sheep today follows a well-defined workflow: remove the wool, throw the fleece onto the wool table, skirt, roll and class the fleece, place it in the appropriate wool bin, press and store the wool until it is transported.

In 1984 Australia became the last country in the world to permit the use of wide combs, due to previous Australian Workers Union rules. Although they were rare in sheds, women now take a large part in the shearing industry by working as pressers, wool rollers, rouse about, wool classers and also shearing, too.

REMOVING THE WOOL

A sheep is caught by the shearer from the catching pen and taken to his "stand" on the shearing board. It is then shorn using mechanical hand piece (see *Shearing devices* below). The wool is removed by following an efficient set of movements, devised by Godfrey Bowen in c. 1950, (the *Bowen Technique*) or the *Tally-Hi* method.

In 1963 the Tally-hi shearing system was developed by the Australian Wool Corporation and promoted using synchronized shearing demonstrations. Sheep struggle less using the Tally-Hi method, reducing strain on the shearer and there is a saving of about 30 seconds shearing each sheep. The shearer begins by removing the sheep's belly wool, which is separated from the main fleece by a rouseabout, while the sheep is still being shorn.

A professional or "gun" shearer typically removes a fleece without badly marking or cutting the sheep in two to three minutes, depending on the size and condition of the sheep, or less than two in elite competitive shearing. The shorn sheep is moved from the board via a chute in the floor, or wall, to a counting out pen, efficiently removing it from the shed.

The CSIRO in Australia has developed a non-mechanical method of shearing sheep using an injected protein that creates a natural break in the wool fibres. After fitting a retaining net to enclose the wool, sheep are injected with the protein. When the net is removed after a week, the fleece has separated and is removed by hand. In some breeds a similar process occurs naturally (see below).

SKIRTING THE FLEECE

Once the entire fleece has been removed from the sheep, the fleece is *thrown*, clean side down, on to a wool table by a shed hand (commonly known in New Zealand and Australian sheds as a *rouseabout* or *roustie*). The wool table top consists of slats spaced approximately 12 cm apart. This enables short pieces of wool, the *locks* and other debris, to gather beneath the table separately from the fleece.

The fleece is then *skirted* by one or more wool rollers to remove the sweat fribs and other less desirable parts of the fleece. The removed pieces largely consist of shorter, seeded, burry or dusty wool etc. which is still useful in the industry. As such they are placed in separate containers and sold along with fleece wool. Other items removed from the fleece on the table, such as faeces, skin fragments or twigs and leaves, are discarded a short distance from the wool table so as not to contaminate the wool and fleece.

WOOL CLASSING

Following the skirting of the fleece, it is folded, rolled and examined for its quality in a process known as wool classing, which is performed by a registered and qualified wool classer. Based on its type, the fleece is placed into the relevant wool bin ready to be pressed (mechanically compressed) when there is sufficient wool to make a wool bale.

SHEARING DEVICES

Blade Shears

Blade shears consist of two blades arranged similarly to scissors except that the hinge is at the end farthest from the point (not in the middle). The cutting edges pass each other as the shearer squeezes them together and shear the wool close to the animal's skin. Blade shears are still used today but in a more limited way. Blade shears leave some wool on a sheep and this is more suitable for cold climates where the sheep needs some protection from the elements. For those areas where no powered-machinery is available blade shears are the only option. Blades are more commonly used to shear stud rams.

Machine Shears

Machine shears, known as hand pieces, operate in a similar manner to human hair clippers in that a power-driven toothed blade, known as a cutter, is driven back and forth over the surface of a comb and the wool is cut from the animal. The original machine shears were powered by a fixed hand-crank linked to the hand piece by a shaft with only two universal joints, which afforded a very limited range of motion. Later models have more joints to allow easier positioning of the hand piece on the animal.

Electric motors on each stand have generally replaced overhead gear for driving the hand pieces. The jointed arm is replaced in many instances with a flexible shaft. Smaller motors allowed the production of shears in which the motor is in the hand piece; these are generally not used by professional shearers as the weight and heat of the motor becomes bothersome with long use.

SHEARING IN AUSTRALASIAN CULTURE

A culture has evolved out of the practice of sheep shearing, especially in post-colonial Australia and New Zealand. *Shearing the Rams*, a painting by

Australian impressionist painter Tom Roberts is considered to be iconic of the livestock-growing culture or "life on the land" in Australia.

For an inversion, Michael Leunig's *Ramming the Shears* can be seen as a sign of the shifts in Australian culture, and the extent to which the dominant rural culture is being eroded by an increasingly urban population.

The expression that Australia's wealth rode on *the sheep's back* in parts of the twentieth century no longer has the currency it once had. In 2001, Mandy Francis of Hardy's Bay, NSW constructed a black butt seat for the Street Furniture Project at Walcha, New South Wales. This seat was inspired by the combs, cutters, wool tables and grating associated with the craft and industry of shearing. During the long weekend in June 2010, 111 machine shearers and 78 blade shearers shore 6,000 Merino ewes and 178 rams at the historic 72 stand *North Tuppal* station. Along with the shearers there were 107 wool handlers and penners-up and more than 10,000 visitors to witness this event in the restored shed. Over this weekend the scene in Tom Robert's *Shearing of the Rams* was twice re-enacted for the visitors.

Many stations across Australia no longer carry sheep due to lower wool prices, drought and other disasters, but their shearing sheds remain, in a wide variety of materials and styles, and have been the subject of books and documentation for heritage authorities. Some farmers are reluctant to remove either the equipment or the sheds, and many unused sheds remain intact.

CONTESTS

Sheep shearing and wool handling competitions are held regularly in parts of the world, particularly Ireland, the UK, South Africa, New Zealand and Australia. As sheep shearing is an arduous task, speed shearers, for all types of equipment and sheep, are usually very fit and well trained. In Wales a sheep shearing contest is one of the events of the Royal Welsh Show, the country's premier agricultural show held near Builth Wells. The world's largest sheep shearing and wool handling contest, the Golden Shears, is held in the Wairarapa district, New Zealand.

SCOURING

Wool straight off a sheep, known as "greasy wool" or "wool in the grease", contains a high level of valuable lanolin, as well as dirt, dead skin, sweat residue, pesticide, and vegetable matter. Before the wool can be used for commercial purposes, it must be scoured, a process of cleaning the greasy wool. Scouring may be as simple as a bath in warm water, or as complicated as an industrial process using detergent and alkali, and specialized equipment.

In commercial wool, vegetable matter is often removed by chemical carbonization. In less processed wools, vegetable matter may be removed by hand, and some of the lanolin left intact through use of gentler detergents. This semi-grease wool can be worked into yarn and knitted into particularly water-resistant mittens or sweaters, such as those of the Aran Island fishermen. Lanolin removed from wool is widely used in cosmetic products such as hand creams.

QUALITY

The quality of wool is determined by the following factors, fibre diameter, crimp, yield, colour, and staple strength. Fibre diameter is the single most important wool characteristic determining quality and price.

Merino wool is typically 3-5 inches in length and is very fine (between 12-24 microns). The finest and most valuable wool comes from Merino hoggets. Wool taken from sheep produced for meat is typically more coarse, and has fibres that are 1.5 to 6 inches in length. Damage or breaks in the wool can occur if the sheep is stressed while it is growing its fleece, resulting in a thin spot where the fleece is likely to break.

Wool is also separated into grades based on the measurement of the wool's diameter in microns and also its style. These grades may vary depending on the breed or purpose of the wool.

For example:

- <15.5 - Ultrafine Merino
- 15.6-18.5 - Superfine Merino
- 18.6-20 - Fine Merino
- 20.1-23 - Medium Merino
- 23< - Strong Merino
- *Comeback*: 21-26 microns, white, 90–180 mm long
- Fine crossbred: 27-31 microns, Corriedales etc.
- Medium crossbred: 32–35 microns
- *Downs*: 23-34 microns, typically lacks luster and brightness. Examples, Aussiedown, Dorset Horn, Suffolk etc.
- *Coarse crossbred*: 36> microns
- *Carpet wools*: 35-45 microns[8]

Any wool finer than 25 microns can be used for garments, while coarser grades are used for outerwear or rugs. The finer the wool, the softer it is, while coarser grades are more durable and less prone to pilling.

The finest Australian and New Zealand Merino wools are known as 1PP which is the industry benchmark of excellence for Merino wool that is 16.9 micron and finer. This style represents the top level of fineness, character, colour, and style as determined on the basis of a series of parameters in accordance with the original dictates of British Wool as applied today by the Australian Wool Exchange (AWEX) Council. Only a few dozen of the millions of bales auctioned every year can be classified and marked 1PP.[14]

HISTORY

As the raw material has been readily available since the widespread domestication of sheep - and of goats, another major provider of wool - the use of felted or woven wool for clothing and other fabrics characterizes some of the earliest civilizations. Prior to invention of shears - probably in the Iron Age - the wool was plucked out by hand or by bronze combs. The oldest known European wool textile, ca. 1500 BCE, was preserved in a Danish bog.[15] Wool fibres from wild goats found in a prehistoric cave in the Republic of Georgia as far back 34,000 BCE suggest that wool fabrics were made even earlier than this.

In Roman times, wool, linen, and leather clothed the European population; the cotton of India was a curiosity that only naturalists had heard of; and silk, imported along the Silk Road from China, was an extravagant luxury. Pliny the Elder records in his Natural History that the reputation for producing the finest wool was enjoyed by Tarentum, where selective breeding had produced sheep with a superior fleece, but which required special care.

In medieval times, as trade connections expanded, the Champagne fairs revolved around the production of wool cloth in small centers such as Proving; the network that the sequence of annual fairs developed meant that the woollens of Provins might find their way to Naples, Sicily, Cyprus, Majorca, Spain, and even Constantinople[18]. The wool trade developed into serious business, the generator of capital.

In the thirteenth century, the wool trade was the economic engine of the Low Countries and of Central Italy; by the end of the following century Italy predominated, though in the 16th century Italian production turned to silk. Both pre-industries were based on English raw wool exports - rivaled only by the sheepwalks of Castile, developed from the fifteenth century - which were a significant source of income to the English crown, which from 1275 imposed an export tax on wool called the "Great Custom".

The importance of wool to the English economy can be shown by the fact that since the 14th Century, the presiding officer of the House of Lords has sat on the "Woolsack", a chair stuffed with wool. Economies of scale were instituted in the Cistercian houses, which had accumulated great tracts of land during the twelfth and early thirteenth centuries, when land prices were low and labour still scarce.

Raw wool was baled and shipped from North Sea ports to the textile cities of Flanders, notably Ypres and Ghent, where it was dyed and worked up as cloth. At the time of the Black Death, English textile industries accounted for about 10 per cent of English wool production; the English textile trade grew during the fifteenth century, to the point where export of wool was discouraged.

Over the centuries, various British laws controlled the wool trade or required the use of wool even in burials. The smuggling of wool out of the country, known as owling, was at one time punishable by the cutting off of a hand. After the Restoration, fine English woollens began to compete with silks in the international market, partly aided by the Navigation Acts; in 1699 English crown forbade its American colonies to trade wool with anyone but England herself.

A great deal of the value of woolen textiles was in the dyeing and finishing of the woven product. In each of the centers of the textile trade, the manufacturing process came to be subdivided into a collection of trades, overseen by an entrepreneur in a system called by the English the "putting-out" system, or "cottage industry", and the Verlagssystem by the Germans.

In this system of producing wool cloth, until recently perpetuated in the production of Harris tweeds, the entrepreneur provides the raw materials and an advance, the remainder being paid upon delivery of the product. Written contracts bound the artisans to specified terms. Fernand Braudel traces the appearance of the system in the thirteenth-century economic boom, quoting a document of 1275[18] The system effectively by-passed the guilds' restrictions.

Before the flowering of the Renaissance, the Medici and other great banking houses of Florence had built their wealth and banking system on their textile industry based on wool, overseen by the Arte della Lana, the wool guild: wool textile interests guided Florentine policies. Francesco Datini, the "merchant of Prato", established in 1383 an Arte della Lana for that small Tuscan city.

The sheepwalks of Castile shaped the landscape and the fortunes of the meseta that lies in the heart of the Iberian peninsula; in the sixteenth century, a unified Spain allowed export of Merino lambs only with royal permission. The German wool market - based on sheep of Spanish origin - did not overtake British wool until comparatively late.

Australia's colonial economy was based on sheep raising, and the Australian wool trade eventually overtook that of the Germans by 1845, furnishing wool for Bradford, which developed as the heart of industrialized woollens production. A World War I era poster sponsored by the United States Department of Agriculture encouraging children to raise sheep to provide needed war supplies.

Due to decreasing demand with increased use of synthetic fibres, wool production is much less than what it was in the past. The collapse in the price of wool began in late 1966 with a 40 per cent drop; with occasional interruptions, the price has tended down. The result has been sharply reduced production and movement of resources into production of other commodities, in the case of sheep growers, to production of meat.

Superwash wool (or washable wool) technology first appeared in the early 1970s to produce wool that has been specially treated so that it is machine washable and may be tumble-dried. This wool is produced using an acid bath that removes the "scales" from the fibre, or by coating the fibre with a polymer that prevents the scales from attaching to each other and causing shrinkage. This process results in a fibre that holds longevity and durability over synthetic materials, while retaining its shape.

In December 2004, a bale of the world's finest wool, averaging 11.8 micron, sold for $3,000 per kilogram at auction in Melbourne, Victoria. This fleece wool tested with an average yield of 74.5 per cent , 68 mm long, and had 40 newtons per kilotex strength. The result was $AUD279,000 for the bale.[24] The finest bale of wool ever auctioned sold for a seasonal record of 269,000 cents per kilo during June 2008. This bale was produced by the Hillcreston Pinehill Partnership and measured 11.6 microns, 72.1 per cent yield and had a 43 Newtons per kilotex strength measurement. The bale realised $247,480 and was exported to India.

During 2007 a new wool suit was developed and sold in Japan that can be washed in the shower, and dries off ready to wear within hours with no ironing required. The suit was developed using Australian Merino wool and it enables woven products made from wool, such as suits, trousers and skirts, to be cleaned using a domestic shower at home. In December 2006 the General Assembly of the United Nations proclaimed 2009 to be the International Year of Natural Fibres, so to raise the profile of wool and other natural fibres.

PRODUCTION

Global wool production is approximately 1.3 million tonnes per year, of which 60 per cent goes into apparel. Australia is the leading producer of wool which is mostly from Merino sheep. New Zealand is the second-largest producer of wool, and the largest producer of crossbred wool. China is the third-largest producer of wool. Breeds such as Lincoln, Romney, Tukidale, Drysdale and Elliotdale produce coarser fibres, and wool from these sheep is usually used for making carpets.

In the United States, Texas, New Mexico and Colorado have large commercial sheep flocks and their mainstay is the Rambouillet (or French Merino). There is also a thriving home-flock contingent of small-scale farmers who raise small hobby flocks of specialty sheep for the hand spinning market. These small-scale farmers offer a wide selection of fleece.

Global woolclip (total amount of wool shorn) 2004/2005:

- Australia: 25 per cent of global woolclip (475 million kg greasy, 2004/2005)
- China: 18 per cent
- New Zealand: 11 per cent
- Argentina: 3 per cent
- Turkey: 2 per cent
- Iran: 2 per cent
- United Kingdom: 2 per cent
- India: 2 per cent
- Sudan: 2 per cent
- South Africa: 1 per cent
- United States: 0.77 per cent

Organic wool is becoming more and more popular. This wool is very limited in supply and much of it comes from New Zealand and Australia. It is becoming easier to find in clothing and other products, but these products often carry a higher price. Wool is environmentally preferable (as compared to petroleum-based Nylon or Polypropylene) as a material for carpets as well, in particular when combined with a natural binding and the use of formaldehyde-free glues. Animal rights groups have noted issues with the production of wool, such as Mulesing.

MARKETING

About 85 per cent of wool sold in Australia is sold by open cry auction. Sale by Sample is a method in which a mechanical claw takes a sample from each bale in a line or lot of wool. These grab samples are bulked, objectively measured, and a sample of not less than 4 kg is displayed in a box for the buyer to examine. The Australian Wool Exchange (AWEX) conducts sales primarily in Sydney, Melbourne, Newcastle, and Fremantle. There are about 80 brokers and agents throughout Australia.

About 7 per cent of Australian wool is sold by private treaty on farms or to local wool-handling facilities. This option gives wool growers benefit from reduced transport, warehousing, and selling costs. This method is preferred for small lots or mixed butts in order to make savings on reclassing and testing. About 5 per cent of Australian wool is sold over the internet on an electronic offer board.

This option gives wool growers the ability to set firm price targets, reoffer passed in wool and offer lots to the market quickly and efficiently. This method works well for tested lots as buyers use these results to make a purchase. 97 per cent of wool is sold without sample inspection however as of dec 2009, 59 per cent of wool listed had been passed in from auction. Growers through certain brokers can allocate their wool to a sale and what price their wool will be reserved at. Sale by tender can achieve considerable cost savings on wool clips large enough to make it worthwhile for potential buyers to submit tenders. Some marketing firms sell wool on a consignment basis, obtaining a fixed percentage as commission.

Forward selling: Some buyers offer a secure price for forward delivery of wool based on estimated measurements or the results of previous clips. Prices are quoted at current market rates and are locked in for the season. Premiums and discounts are added to cover variations in micron, yield, tensile strength, etc., which are confirmed by actual test results when available.Another method of selling wool includes sales direct to wool mills.

The British Wool Marketing Board operates a central marketing system for UK fleece wool with the aim of achieving the best possible net returns for farmers. Less than half of New Zealand's wool is sold at auction, while around 45 per cent for farmers sell wool directly to private buyers and end-users.

Some businesses in New Zealand like Blue House Yarns have turned to selling organic wool, a new trend on wool production.

United States sheep producers' market wool with private or cooperative wool warehouses, but wool pools are common in many states. In some cases, wool is pooled in a local market area but sold through a wool warehouse. Wool offered with objective measurement test results is preferred. Imported apparel wool and carpet wool goes directly to central markets where it is handled by the large merchants and manufacturers.

USES

In addition to clothing, wool has been used for blankets, horse rugs, saddle cloths, carpeting, felt, wool insulation (also see links) and upholstery. Wool felt covers piano hammers, and it is used to absorb odours and noise in heavy machinery and stereo speakers.

Ancient Greeks lined their helmets with felt, and Roman legionnaires used breastplates made of wool felt.

Wool has also been traditionally used to cover cloth diapers. Wool fibre exteriors are hydrophobic (repel water) and the interior of the wool fibre is hygroscopic (attracts water); this makes a wool garment able to cover a wet diaper while inhibiting wicking, so outer garments remain dry. Wool felted and treated with lanolin is water resistant, air permeable, and slightly antibacterial, so it resists the buildup of odour. Some modern cloth diapers use felted wool fabric for covers, and there are several modern commercial knitting patterns for wool diaper covers.

Initial studies of woollen underwear have found it prevented heat and sweat rashes because it more readily absorbs the moisture than other fibres.Merino wool has been used in baby sleep products such as swaddle baby wrap blankets and infant sleeping bags. Wool is an animal protein, and as such it can be used as a soil fertiliser, being a slow release source of nitrogen and ready made amino acids.

VIRGIN WOOL

Wool spun for the first time is called Virgin wool.

SHODDY OR RECYCLED WOOL:

It is made by cutting or tearing apart existing wool fabric and respinning the resulting fibres. As this process makes the wool fibres shorter, the remanufactured fabric is inferior to the original. The recycled wool may be mixed with raw wool, wool noil, or another fibre such as cotton to increase the average fibre length. Such yarns are typically used as weft yarns with a cotton warp. This process was invented in the Heavy Woollen District of West Yorkshire and created a micro-economy in this area for many years.

Ragg is a sturdy wool fibre made into yarn and used in many rugged applications like gloves. Worsted is a strong, long-staple, combed wool yarn with a hard surface.

Woollen is a soft, short-staple, carded wool yarn typically used for knitting. In traditional weaving, woollen weft yarn (for softness and warmth) is frequently combined with a worsted warp yarn for strength on the loom.

EVENTS

A buyer of Merino wool, Ermenegildo Zegna, has offered awards for Australian wool producers. In 1963, the first Ermenegildo Zegna Perpetual Trophy was presented in Tasmania for growers of "Superfine skirted Merino fleece". In 1980, a national award, the Ermenegildo Zegna Trophy for Extrafine Wool Production, was launched. In 2004, this award became known as the Ermenegildo Zegna Unprotected Wool Trophy. In 1998, an Ermenegildo Zegna Protected Wool Trophy was launched for fleece from sheep coated for around nine months of the year.

In 2002, the Ermenegildo Zegna Vellus Aureum Trophy was launched for wool that is 13.9 micron and finer. Wool from Australia, New Zealand, Argentina, and South Africa may enter, and a winner is named from each country. In April 2008, New Zealand won the Ermenegildo Zegna Vellus Aureum Trophy for the first time with a fleece that measured 10.8 microns. This contest awards the winning fleece weight with the same weight in gold as a prize, hence the name.

In 2010 an ultra-fine, 10 micron fleece, from Windradeen, near Pyramul, New South Wales set a new world record in the fineness of wool fleeces when it won the Ermenegildo Zegna Vellus Aureum International Trophy.

Since 2000, Loro Piana has awarded a cup for the world's finest bale of wool that produces just enough fabric for 50 tailor-made suits. The prize is awarded to an Australian or New Zealand wool grower who produces the year's finest bale.

The New England Merino Field days which display local studs, wool, and sheep are held during January, every two years (in even numbered years) around the Walcha, New South Wales district. The Annual Wool Fashion Awards, which showcase the use of Merino wool by fashion designers, are hosted by the city of Armidale, New South Wales in March each year.

This event encourages young and established fashion designers to display their talents. During each May, Armidale hosts the annual New England Wool Expo to display wool fashions, handicrafts, demonstrations, shearing competitions, yard dog trials, and more.

In July, the annual Australian Sheep and Wool Show is held in Bendigo, Victoria. This is the largest sheep and wool show in the world, with goats and alpacas as well as woolcraft competitions and displays, fleece competitions, sheepdog trials, shearing, and wool handling. The largest competition in the world for objectively-measured fleeces is the Australian Fleece Competition, which is held annually at Bendigo. In 2008, there were 475 entries from all states of Australia with first and second prizes going to the Northern Tablelands, New South Wales fleeces.

WOOL CLASSING

Wool classing is an occupation for which people are trained to produce uniform, predictable, low risk lines of wool. This is carried out by examining the characteristics of the wool in its raw state.

The characteristics which a wool classer would examine are:

- *Breed of the sheep*: Shedding breeds will increase the risk of medulated and/or pigmented fibres. Any sheep likely to have dark fibres should be shorn last to avoid contamination. The age of the sheep will have a bearing on the fibre diameter and value of wool, too.
- *Chemical usage*: Ensure that all rules have been followed.
- *Brands*, *seedy jowls* and *shanks*: Must be removed from fleeces and broken.
- *Stain*: Must be removed from bellies and fleeces and identified in a separate line.
- *Wool crimp*: The number of bends per unit length along the wool fibre approximately indicates spinning capacity of the wool. fibres with a fine crimp have many bends and usually have a small diameter. Such fibre can be spun into fine yarns, with great lengths of yarn for a given weight of wool, and greater market value. Fine fibres may be utilised in the production of fine garments such as men's suits whereas the coarser fibres may be used for the production of carpet and other sturdy products. Crimp is measured in crimps per inch or crimps per centimetre. Average diameter or mean fibre diameter is measured in micrometres (microns). For generations, English wool-handlers categorized wool along the above lines estimating spinning capacity by eye and touch. This spread worldwide as the Bradford system.
- *Wool Strength* (also known as *tensile strength*) determines wool's ability to withstand processing. Weaker wools produce more waste in carding and spinning. Weaker wools may be used for production of felt, or combined with other fibres, etc.
- *Wool colour*: Indicates whether wool is able to be dyed in light shades. Colour may be graded depending upon the natural colour, impurities and various stains present. Severely stained wool decreases prices dramatically. However, it is difficult to assess colour accurately without proper measurement, since some stains will wash out in the processing, whereas others are quite persistent.

The fleece is skirted to remove excess frib, seed and burr etc to leave the fleece as reasonably even as possible in good respects. The parts of wool taken from a sheep are graded separately. The fleece forming the bulk of the yield is placed with other fleece wool as the main line, other pieces such as the neck, belly and skirtings (inferior wool from edges) are sold for such purposes where the shorter wools are required (for example: fillings, carpets, insulation).

Whilst in some places crimp may determine which grade the fleece will be placed into, this subjective assessment is not always reliable and processors prefer that wools are measured objectively by qualified laboratories. Some of the superfine wool growers do in shed wool testing, but this can only be used as a guide. This enables wool classers to place wool into lines of a consistent quality.

A shedhand, known as a wool presser, places the wool into approved wool packs in a wool press to produce a bale of wool that must meet regulations concerning its fastenings, length, weight and branding if it is to be sold at auction in Australasia. All Merino fleece wool sold at auction in Australia is objectively measured for fibre diameter, yield (including the amount of vegetable matter), staple length, staple strength and sometimes colour.

Classers are also responsible for ensuring that a pre-shearing check is made to ensure that the wool and sheep areas are free of possible contaminants. They are to supervise shed staff during shearing and train any inexperienced hands. At the end of shearing classers have to provide full documentation concerning the clip.

ROOING

In some primitive sheep (for example in many Shetlands), there is a natural break in the growth of the wool in spring. By late spring this causes the fleece to begin to peel away from the body, and it may then be plucked by hand without cutting – this is known as *rooing*. Individual sheep may reach this stage at slightly different times. Shearing can be done with use of hand-shears or powered shears. Professional sheep shearers can shear a sheep in under a minute, without nicking the sheep.

The fleece is removed in one piece. Second cuts can be made but produce only short fibres, which are more difficult to spin. Primitive breeds, like the Scottish Soay sheep have to be plucked, not sheared, as the kemps are still longer than the soft fleece, (a process called rooing).

SKIRTING

Skirting is disposing of all wool that is unsuitable for spinning. Recovering can be attempted. It can also be done at the same time as carding.

CLEANING

The wool is cleaned. At this point the fleece is full of lanolin and often contains extraneous vegetable matter, such as sticks, twigs, burrs and straw. These may all be removed, though lanolin may be left in the wool till after the spinning, a technique known as spinning 'in the grease'. Indeed if the fabric is to be water repellent, lanolin is not removed at any stage.

Washing the wool at this stage can be a tedious process. Some people wash it a small handful at a time very carefully, and then set it out to dry on a table in the sun. Others will wash the whole fleece. Lanolin is removed by soaking the fleece in very hot water. If the fleece gets agitated, it will become felt, and then spinning is impossible. Felting, when done on purpose (with needles, chemicals, or simply rubbing the fibres against each other), can be used to create garments.

CARDING OR COMBING

It is possible to spin directly from a clean fleece, but it is much easier to spin a carded fleece. Carding by hand yields a rolag, a loose woollen roll of fibres. Using a drum carder yields a bat, which is a mat of fibres in a flat, rectangular shape. Carding mills return the fleece in a roving, which is a stretched bat; it is very long and often the thickness of a wrist.A pencil roving is a roving thinned to the width of a pencil. It can used for knitting without any spinning, or for apprentices.

One good-sized fleece may take weeks to card with a drum-carder, or an eternity by hand. If the fleece is sent to a carding mill, it must be washed before it is carded. Most mills offer washing the wool as a service, with extra fees if the wool is exceptionally dirty. Fibres can be purchased pre-carded. Combing is another method to align the fibres parallel to the yarn, and thus is good for spinning a worsted yarn, whereas the rolag from handcards produces a woolen yarn.

SPINNING

Hand spinning can be done by using a spindle or the spinning wheel. Spinning turns the carded wool fibres into yarn which can then be directly woven, knitted (flat or circular), crocheted, or by other means turned into fabric or a garment. The spinning wheel collects the yarn on a bobbin. A woollen yarn is lightly spun so it is airey, and is a good insulator and suitable for knitting, while a worsted yarn is spun tight to exclude air, and has greater strength and is suited to weaving. Once the bobbin is full, the spinner either puts on a new bobbin, or forms a skein, or balls the yarn. A skein is a coil of yarn twisted into a loose knot. Yarn is skeined using a niddy-noddy or other type of skein -winder. Yarn is rarely balled directly after spinning, it will be stored in skein form, and transferred to a ball only if needed. Knitting from a skein, is difficult as the yarn forms knots, in this case it is best to ball. Yarn to be plied is left on the bobbin.

A skein is either formed on a niddy noddy or some other type of skein winder. Traditionally niddy-noddys looked like an uppercase "i", with the bottom half rotated 90 degrees. Now spinning wheel manufactures also make niddy-noddys that attach onto the spinning wheel for faster skein winding.

PLYING

Plying yarn is when one takes a strand of spun yarn (one strand is often called a single) and spins it together with other strands in order to make a thicker yarn.

Regular plying consists of taking two or more singles and twisting them together, the against their twist. This can be done on a spinning wheel or on a spindle. If the yarn was spun clockwise (which is called a "Z" twist), to ply, the wheel must spin counter-clockwise (an "S" twist). This is the most common way. When plying from bobbins a device called a lazy kate is often used to hold them.

Most spinners (who use spinning wheels) ply from bobbins. This is easier than plying from balls because there is less chance for the yarn to become tangled and knotted if it is simply unwound from the bobbins. So that the bobbins can unwind freely, they are put in a device called a lazy kate, or sometimes simply *kate*.

The simplest lazy kate consists of wooden bars with a metal rod running between them. Most hold between three and four bobbins. The bobbin sits on the metal rod. Other lazy kates are built with devices that create an adjustable amount of tension, so that if the yarn is jerked, a whole bunch of yarn is not wound off, then wound up again in the opposite direction. Some spinning wheels come with a built in lazy kate. Navajo plying consists of making large loops, similar to crocheting.A loop about 8 inches long is made on the leader the end on the leader.

(A leader is the string left on the bobbin to spin off.) The three strands together are spun in the opposite direction. When a third of the loop remains, a new loop is created and the spinning continues.

The process is repeated until the yarn is all plied. The advantage of this method is that only one single is needed and if the single is already dyed this technique allows it to be plied without ruining the colour scheme. This technique also allows the spinner to try to match up thick and thin spots in the yarn, thus making for a smoother end product.

WASHING

If the lanolin is unwanted, and has not already been washed out, this is done now. The skein is tied in six points and steeped overnight in detergent, it is rinsed and air-dried, and re-skinned.

Unless the lanolin is to be left in the cloth as a water repellent. When washing a skein it works well to let the wool soak in soapy water overnight, and rinse the soap out in the morning. Dishwashing detergents are commonly used, and a special laundry detergent designed for washing wool is not required.

The dishwashing detergent works and does not harm the wool. After washing, let the wool dry (air drying works best). Once it is dry, or just a bit damp, one can stretch it out a bit on a niddy-noddy.

Putting the wool back on the niddy-noddy makes for a nicer looking finished skein. Before taking a skein and washing it, the skein must be tied up loosely in about six places. If the skein is not tied up, it will be very hard to unravel when done washing.

FLAX

The preparations for spinning is similar across most plant fibres, including Flax and Hemp. Flax is the fibre used to create linen. Cotton is handled differently since it uses the fruit of the plant and not the stalk.

HARVESTING

Flax is pulled out of the ground about a month after the initial blooming when the lower part of the plant begins to turn yellow, and when the most forward of the seeds are found in a soft state. It is pulled in handfuls and several handfuls are tied together with slip knot into a 'beet'. The string is tightened as the stalks dry. The seed heads are removed and the seeds collected, by threshing and winnowing.

RETTING

Threshing and dressing flax at the Roscheider Hof Open Air Museum Retting is the process of rotting away the inner stalk, leaving the outer fibres intact. A standing pool of warm water is needed, into which the beets are submerged. An acid is produced when retting, and it would corrode a metal container. At 80 °F (27 °C), the retting process takes 4 or 5 days, it takes longer takes longer when colder. When the retting is complete the bundles feel soft and slimy, The process can be overdone, and the fibres rot too.

DRESSING THE FLAX

Dressing is removing the fibres from the straw and cleaning it enough to be spun. The flax is broken, scutched and hackled in this step. Breaking The process of breaking breaks up the straw into short segments. The beets are untied and fed between the beater of the breaking machine, the set of wooden blades that mesh together when the upper jaw is lowered.

Scutching In order to remove some of the straw from the fibre a wooden scutching knife is scaped down the fibres while they hang vertically. Heckling Fibre is pulled through various different sized heckling combs. A Heckling comb is a bed of sharp, long-tapered, tempered, polished steel pins driven into wooden blocks at regular spacing. A good progression is from 4 pins per square inch, to 12, to 25 to 48 to 80. The first three will remove the straw, and the last two will split and polish the fibres. Some of the finer stuff that comes off in the last heckles can be carded like wool and spun. It will produce a coarser yarn than the fibres pulled through the heckles because it will still contain some straw.

SPINNING THE FLAX

Flax can either be spun from a distaff, or from the spinner's lap. Spinners keep their fingers wet when spinning, to prevent forming fuzzy thread. Usually singles are spun with an "S" twist. After flax is spun it is washed in a pot of boiling water for a couple of hours to set the twist and reduce fuzziness.

Many handspinners, will buy a roving of flax. This roving is spun in the same manner as above. The rovings may come with very long fibres (4 to 8 inches), or much shorter fibres (2 to 3 inches).

SILK

Silk is a natural protein fibre, some forms of which can be woven into textiles. The best-known type of silk is obtained from the cocoons of the larvae of the mulberry silkworm *Bombyx mori* reared in captivity (sericulture). The shimmering appearance of silk is due to the triangular prism-like structure of the silk fibre, which allows silk cloth to refract incoming light at different angles, thus producing different colours.

Silks are produced by several other insects, but only the silk of moth caterpillars has been used for textile manufacturing. There has been some research into other silks, which differ at the molecular level. Silks are mainly produced by the larvae of insects undergoing complete metamorphosis, but also by some adult insects such as webspinners. Silk production is especially common in the Hymenoptera (bees, wasps, and ants), and is sometimes used in nest construction. Other types of arthropod produce silk, most notably various arachnids such as spiders (see spider silk).

HISTORY

Woven silk textile from tomb no 1. at Mawangdui in Changsha, Hunan province, China, from the Western Han Dynasty, 2nd century BC

WILD SILK

A variety of wild silks, produced by caterpillars other than the mulberry silkworm have been known and used in China, South Asia, and Europe since ancient times. However, the scale of production was always far smaller than that of cultivated silks. They differ from the domesticated varieties in colour and texture, and cocoons gathered in the wild usually have been damaged by the emerging moth before the cocoons are gathered, so the silk thread that makes up the cocoon has been torn into shorter lengths.

Commercially reared silkworm pupae are killed by dipping them in boiling water before the adult moths emerge, or by piercing them with a needle, allowing the whole cocoon to be unravelled as one continuous thread. This permits a much stronger cloth to be woven from the silk. Wild silks also tend to be more difficult to dye than silk from the cultivated silkworm.

CHINA

Silk fabric was first developed in ancient China, with some of the earliest examples found as early as 3500 BC. Legend gives credit for developing silk to a Chinese empress, Lei Zu (Hsi-Ling-Shih, Lei-Tzu). Silks were originally reserved for the Kings of China for their own use and gifts to others, but spread gradually through Chinese culture and trade both geographically and socially, and then to many regions of Asia.

Silk rapidly became a popular luxury fabric in the many areas accessible to Chinese merchants because of its texture and luster. Silk was in great demand, and became a staple of pre-industrial international trade. In July 2007, archeologists discovered intricately woven and dyed silk textiles in a tomb in Jiangxi province, dated to the Eastern Zhou Dynasty roughly 2,500 years ago.

Although historians have suspected a long history of a formative textile industry in ancient China, this find of silk textiles employing "complicated techniques" of weaving and dyeing provides direct and concrete evidence for silks dating before the Mawangdui-discovery and other silks dating to the Han Dynasty (202 BC-220 AD).

The first evidence of the silk trade is the finding of silk in the hair of an Egyptian mummy of the 21st dynasty, c.1070 BC. Ultimately the silk trade reached as far as the Indian subcontinent, the Middle East, Europe, and North Africa. This trade was so extensive that the major set of trade routes between Europe and Asia has become known as the Silk Road. The highest development was in China.

The Emperors of China strove to keep knowledge of sericulture secret to maintain the Chinese monopoly. Non-etheless sericulture reached Korea around 200 BC, about the first half of the 1st century AD had reached ancient Khotan, and by AD 300 the practice had been established in India.

THAILAND

Silk is produced, year round, in Thailand by two types of silkworms, the cultured Bombycidae and wild Saturniidae. Most production is after the rice harvest in the southern and northeast parts of the country. Women traditionally weave silk on hand looms, and pass the skill on to their daughters as weaving is considered to be a sign of maturity and eligibility for marriage. Thai silk textiles often use complicated patterns in various colours and styles. Most regions of Thailand have their own typical silks.

A single thread filament is too thin to use on its own so women combine many threads to produce a thicker, usable fibre. They do this by hand-reeling the threads onto a wooden spindle to produce a uniform strand of raw silk. The process takes around 40 hours to produce a half kilogram of Thai silk.

Many local operations use a reeling machine for this task, but some silk threads are still hand-reeled. The difference is that hand-reeled threads produce three grades of silk: two fine grades that are ideal for lightweight fabrics, and a thick grade for heavier material.

The silk fabric is soaked in extremely cold water and bleached before dyeing to remove the natural yellow colouring of Thai silk yarn. To do this, skeins of silk thread are immersed in large tubs of hydrogen peroxide. Once washed and dried, the silk is woven on a traditional hand operated loom.

INDIA

Silk, known as "Paat" in Eastern India, *Pattu* in southern parts of India and *Resham* in Hindi/Urdu, has a long history in India. Recent archaeological discoveries in Harappa and Chanhu-daro suggest that sericulture, employing wild silk threads from native silkworm species, existed in South Asia during the time of the Indus Valley Civilization, roughly contemporaneous with the earliest known silk use in China. Silk is widely produced today.

India is the second largest producer of silk after China. A majority of the silk in India is produced in Karnataka State, particularly in Mysore and the North Bangalore regions of Muddenahalli, Kanivenarayanapura, and Doddaballapur. India is also the largest consumer of silk in the world. The tradition of wearing silk sarees in marriages by the brides is followed in southern parts of India. Silk is worn by people as a symbol of royalty while attending functions and during festivals.

Historically silk was used by the upper classes, while cotton was used by the poorer classes. Today silk is mainly produced in Bhoodhan Pochampally (also known as Silk City), Kanchipuram, Dharmavaram, Mysore, etc. in South India and Banaras in the North for manufacturing garments and sarees. "Murshidabad silk", famous from historical times, is mainly produced in Malda and Murshidabad district of West Bengal and woven with hand looms in Birbhum and Murshidabad district. Another place famous for production of silk is Bhagalpur.

The silk from Pochampally is particularly well-known for its classic designs and enduring quality. The silk is traditionally hand-woven and hand-dyed and usually also has silver threads woven into the cloth. Most of this silk is used to make sarees. The sarees usually are very expensive and vibrant in colour. Garments made from silk form an integral part of Indian weddings and other celebrations.

In the northeastern state of Assam, three different types of silk are produced, collectively called Assam silk: Muga, Eri and Pat silk. Muga, the golden silk, and Eri are produced by silkworms that are native only to Assam. The heritage of silk rearing and weaving is very old and continues today especially with the production of Muga and Pat *riha* and *mekhela chador*, the three-piece silk sarees woven with traditional motifs. *Mysore Silk Sarees*, which are known for their soft texture, last many years if carefully maintained.

ANCIENT MEDITERRANEAN

In the Odyssey, 19.233, when Odysseus, while pretending to be someone else, is questioned by Penelope about her husband's clothing, he says that he wore a shirt "gleaming like the skin of a dried onion" (varies with translations, literal translation here) which could refer to the lustrous quality of silk fabric. The Roman Empire knew of and traded in silk. During the reign of emperor Tiberius, sumptuary laws were passed that forbade men from wearing silk garments, but these proved ineffectual.

Despite the popularity of silk, the secret of silk-making only reached Europe around AD 550, via the Byzantine Empire. Legend has it those monks working for the emperor Justinian I smuggled silkworm eggs to Constantinople in hollow canes from China. All top-quality looms and weavers were located inside the Palace complex in Constantinople and the cloth produced was used in imperial robes or in diplomacy, as gifts to foreign dignitaries. The remainder was sold at very high prices.

MIDDLE EAST

In Islamic teachings, Muslim men are forbidden to wear silk. Many religious jurists believe the reasoning behind the prohibition lies in avoiding clothing for men that can be considered feminine or extravagant. There are disputes regarding the amount of silk a fabric can consist of (*i.e.*, whether a small decorative silk piece on a cotton caftan is permissible or not) for it to be lawful for men to wear but the dominant opinion of most Muslim scholars is that the wearing of silk for men is forbidden.

Despite injunctions against silk for men, silk has retained its popularity in the Islamic world because of its permissibility for women. The Muslim Moors brought silk with them to Spain during their conquest of the Iberian Peninsula.

MEDIEVAL AND MODERN EUROPE

Venetian merchants traded extensively in silk and encouraged silk growers to settle in Italy. By the 13th century, Italian silk was a significant source of trade. Since that period, the silk worked in the province of Como has been the most valuable silk in the world. The wealth of Florence was largely built on textiles, both wool and silk and other cities like Lucca also grew rich on the trade.

Italian silk was so popular in Europe that Francis I of France invited Armenian silk makers to France to create a French silk industry, especially in

Lyon. Mass emigration (especially of Huguenots) during periods of religious dispute had seriously damaged French industry and introduced these various textile industries, including silk, to other countries.

James I attempted to establish silk production in England, purchasing and planting 100,000 mulberry trees, some on land adjacent to Hampton Court Palace, but they were of a species unsuited to the silk worms, and the attempt failed. British enterprise also established silk filature in Cyprus in 1928. In England in the mid 20th Century, silk was produced at Lullingstone Castle in Kent. Silkworms were raised and reeled under the direction of Zoe Lady Hart Dyke. Production started elsewhere later. In Italy, the Stazione Bacologica Sperimentale was founded in Padua in 1871 to research sericulture. In the late 19th century, China, Japan, and Italy were the major producers of silk.

The most important cities for silk production in Italy were Como and Meldola (Forlì). In medieval times, it was common for silk to be used to make elaborate casings for bananas and other fruits. Silk was expensive in Medieval Europe and used only by the rich. Italian merchants like Giovanni Arnolfini became hugely wealthy trading it to the Courts of Northern Europe.

NORTH AMERICA

James I of England introduced silk-growing to the American colonies around 1619, ostensibly to discourage tobacco planting. The Shakers in Kentucky adopted the practice as did a cottage industry in New England. In the 19th century a new attempt at a silk industry began with European-born workers in Paterson, New Jersey, and the city became a US silk center, although Japanese imports were still more important.

World War II interrupted the silk trade from Japan. Silk prices increased dramatically, and US industry began to look for substitutes, which led to the use of synthetics such as nylon. Synthetic silks have also been made from local, a type of cellulose fibre, and are often difficult to distinguish from real silk (see spider silk for more on synthetic silks).

PROPERTIES

Physical Properties

Silk fibres from the *Bombyx mori* silkworm have a triangular cross section with rounded corners, 5-10 ìm wide. The fibroin-heavy chain is composed mostly of beta-sheets, due to a 59-mer amino acid repeat sequence with some variations. The flat surfaces of the fibrils reflect light at many angles, giving silk a natural shine. The cross-section from other silkworms can vary in shape and diameter: crescent-like for *Anaphe* and elongated wedge for *tussah*.

Silkworm fibres are naturally extruded from two silkworm glands as a pair of primary filaments (brin), which are stuck together, with sericin proteins that act like glue, to form a bave. Bave diameters for tussah silk can reach 65 ìm.

See cited reference for cross-sectional SEM photographs. Silk has a smooth, soft texture that is not slippery, unlike many synthetic fibres. Silk is one of the strongest natural fibres but loses up to 20 per cent of its strength when wet. It has a good moisture regain of 11 per cent . Its elasticity is moderate to poor: if elongated even a small amount, it remains stretched. It can be weakened if exposed to too much sunlight. It may also be attacked by insects, especially if left dirty.

Silk is a poor conductor of electricity and thus susceptible to static cling. Unwashed silk chiffon may shrink up to 8 per cent due to a relaxation of the fibre macrostructure. So silk should either be pre-washed prior to garment construction, or dry cleaned. Dry cleaning may still shrink the chiffon up to 4 per cent . Occasionally, this shrinkage can be reversed by a gentle steaming with a press cloth. There is almost no gradual shrinkage nor shrinkage due to molecular-level deformation.

Natural and synthetic silk is known to manifest piezoelectric properties in proteins, probably due to its molecular structure.Silkworm silk was used as the standard for the denier, a measurement of linear density in fibres. Silkworm silk therefore has a linear density of approximately 1 den, or 1.1 dtex.

Comparison ofLinear	Diameter (μm)	Coeff. Variation	Silk Fibres Density(dtex)
Moth: *Bombyx mori*	1.17	12.9	24.8%
Spider: *Argiope aurentia*	0.14	3.57	14.8%

Chemical properties

Silk emitted by the silkworm consists of two main proteins, sericin and fibroin, fibroin being the structural center of the silk, and serecin being the sticky material surrounding it. Fibroin is made up of the amino acids Gly-Ser-Gly-Ala-Gly-Ala and forms beta pleated sheets. Hydrogen bonds form between chains, and side chains form above and below the plane of the hydrogen bond network.

The high proportion (50 per cent) of glycine, which is a small amino acid, allows tight packing and the fibres are strong and resistant to stretching. The tensile strength is due to the many interseeded hydrogen bonds. Since the protein forms a beta sheet, when stretched the force is applied to these strong bonds and they do not break.

Silk is resistant to most mineral acids, except for sulfuric acid, which dissolves it. It is yellowed by perspiration.

USES

Silk's absorbency makes it comfortable to wear in warm weather and while active. Its low conductivity keeps warm air close to the skin during cold weather. It is often used for clothing such as shirts, ties, blouses, formal dresses, high fashion clothes, lingerie, pyjamas, robes, dress suits, sun dresses and kimonos.

Silk's attractive luster and drape makes it suitable for many furnishing applications. It is used for upholstery, wall coverings, window treatments (if blended with another fibre), rugs, bedding and wall hangings

While on the decline now, due to artificial fibres, silk has had many industrial and commercial uses; parachutes, bicycle tires, comforter filling and artillery gunpowder bags. A special manufacturing process removes the outer irritant sericin coating of the silk, which makes it suitable as non-absorbable surgical sutures. This process has also recently led to the introduction of specialist silk underclothing for children and adults with eczema where it can significantly reduce itch.

PRODUCTION

The cultivation of silk is called sericulture. Over 30 countries produce silk, the major ones are China (54 per cent) and India (14 per cent).

To produce 1 kg of silk, 104 kg of mulberry leaves must be eaten by 3000 silkworms. It takes about 5000 silkworms to make a pure silk kimono.

Table. Top Ten Cocoons (Reelable) Producers — 2005

Country	Production (Int $1000)	Footnote	Production (1000 kg)	Footnote
People's Republic of China	978,013	C	290,003	F
India	259,679	C	77,000	F
Uzbekistan	57,332	C	17,000	F
Brazil	37,097	C	11,000	F
Iran	20,235	C	6,000	F
Thailand	16,862	C	5,000	F
Vietnam	10,117	C	3,000	F
Democratic People's Republic of Korea	5,059	C	1,500	F
Romania	3,372	C	1,000	F
Japan	2,023	C	600	F

Note: No symbol = official figure
F = FAO estimate
* = Unofficial figure
C = Calculated figure

Production in Int $1000 have been calculated based on 1999-2001 international pricesSource: Food And Agricultural Organization of United Nations: Economic And Social Department: The Statistical Division

CULTIVATION

Silk moths lay eggs on specially prepared paper. The eggs hatch and the caterpillars (silkworms) are fed fresh mulberry leaves. After about 35 days and

4 moltings, the caterpillars are 10,000 times heavier than when hatched and are ready to begin spinning a cocoon. A straw frame is placed over the tray of caterpillars, and each caterpillar begins spinning a cocoon by moving its head in a pattern.

Two glands produce liquid silk and force it through openings in the head called spinnerets. Liquid silk is coated in sericin, a water-soluble protective gum, and solidifies on contact with the air. Within 2–3 days, the caterpillar spins about 1 mile of filament and is completely encased in a cocoon. The silk farmers then kill most caterpillars by heat, leaving some to metamorphose into moths to breed the next generation of caterpillars.

Harvested cocoons are then soaked in boiling water to soften the sericin holding the silk fibres together in a cocoon shape. The fibres are then unwound to produce a continuous thread. Since a single thread is too fine and fragile for commercial use, anywhere from three to ten strands are spun together to form a single thread of silk.

ANIMAL RIGHTS

As the process of harvesting the silk from the cocoon kills the larvae, sericulture has been criticized in the early 21st century by animal rights activists, especially since artificial silks are available. Mohandas Gandhi was also critical of silk production based on the Ahimsa philosophy "not to hurt any living thing." This led to Gandhi's promotion of cotton spinning machines, an example of which can be seen at the Gandhi Institute. He also promoted *Ahimsa silk*, wild silk made from the cocoons of wild and semi-wild silk moths. Ahimsa silk is promoted in parts of Southern India for those who prefer not to wear silk produced by killing silkworms.

COTTON

Cotton is a soft, fluffy staple fibre that grows in a boll, or protective capsule, around the seeds of cotton plants of the genus *Gossypium*. The plant is a shrub native to tropical and subtropical regions around the world, including the Americas, Africa, India, and Pakistan. The fibre most often is spun into yarn or thread and used to make a soft, breathable textile, which is the most widely used natural-fibre cloth in clothing today. The English name derives from the Arabic *(al) qutn*, which began to be used circa 1400. The botanical purpose of cotton fibre is to aid in seed dispersal.

HISTORY

According to the Foods and Nutrition Encyclopaedia, the earliest cultivation of cotton discovered thus far in the Americas occurred in Mexico, some 8,000 years ago. The indigenous species was *Gossypium hirsutum*, which is today the most widely planted species of cotton in the world, constituting about 89.9 per cent of all production worldwide. The greatest diversity of wild cotton species is found in Mexico, followed by Australia and Africa. Cotton was first cultivated in the Old World 7,000 years ago (5th–4th millennia BC), by the inhabitants of the Indus Valley Civilization, which covered a huge swath of the northwestern part of the Indian subcontinent, comprising today parts of eastern Pakistan and northwestern India. The Indus cotton industry was well developed and some methods used in cotton spinning and fabrication continued to be used until the modern industrialization of India. Well before the Common Era, the use of cotton textiles had spread from India to the Mediterranean and beyond.

Greeks and the Arabs were apparently ignorant about cotton until the Wars of Alexander the Great, as his contemporary Megasthenes told Seleucus I Nicator of "there being trees on which wool grows" in "Indica".

According to The Columbia Encyclopaedia, Sixth Edition:

Cotton has been spun, woven, and dyed since prehistoric times. It clothed the people of ancient India, Egypt, and China. Hundreds of years before the Christian era, cotton textiles were woven in India with matchless skill, and their use spread to the Mediterranean countries. In the first century, Arab traders brought fine muslin and calico to Italy and Spain.

The Moors introduced the cultivation of cotton into Spain in the 9th century. Fustians and dimities were woven there and in the 14th century in Venice and Milan, at first with a linen warp. Little cotton cloth was imported to England before the 15th century, although small amounts were obtained chiefly for candlewicks. By the 17th century, the East India Company was bringing rare fabrics from India. Native Americans skillfully spun and wove cotton into fine garments and dyed tapestries. Cotton fabrics found in Peruvian tombs are said to belong to a pre-Inca culture.

In Iran (Persia), the history of cotton dates back to the Achaemenid era (5th century BC); however, there are few sources about the planting of cotton in pre-Islamic Iran. The planting of cotton was common in Merv, Ray and Pars of Iran. In the poems of Persian poets, especially Ferdowsi's Shahname, there are many references to cotton ("panbe" in Persian). Marco Polo (13th century) refers to the major products of Persia, including cotton. John Chardin, a famous French traveller of 17th century, who had visited the Safavid Persia, has approved the vast cotton farms of Persia.

In Peru, cultivation of the indigenous cotton species *Gossypium barbadense* was the backbone of the development of coastal cultures, such as the Norte Chico, Moche and Nazca. Cotton was grown upriver, made into nets and traded with fishing villages along the coast for large supplies of fish. The Spanish who came to Mexico and Peru in the early 16th century found the people growing cotton and wearing clothing made of it.

During the late medieval period, cotton became known as an imported fibre in northern Europe, without any knowledge of how it was derived, other than that it was a plant; noting its similarities to wool, people in the region could only imagine that cotton must be produced by plant-borne sheep. John Mandeville, writing in 1350, stated as fact the now-preposterous belief: "There grew there [India] a wonderful tree which bore tiny lambs on the endes of its branches.

These branches were so pliable that they bent down to allow the lambs to feed when they are hungrie [*sic*]." (See Vegetable Lamb of Tartary.) This aspect is retained in the name for cotton in many European languages, such as German *Baumwolle*, which translates as "tree wool" (*Baum* means "tree"; *Wolle* means "wool"). By the end of the 16th century, cotton was cultivated throughout the warmer regions in Asia and the Americas.

India's cotton-processing sector gradually declined during British expansion in India and the establishment of colonial rule during the late 18th and early 19th centuries. This was largely due to aggressive colonialist mercantile policies of the British East India Company, which made cotton processing and manufacturing workshops in India uncompetitive. Indian markets were increasingly forced to supply only raw cotton and were forced, by British-imposed law, to purchase manufactured textiles from Britain.

INDUSTRIAL REVOLUTION IN BRITAIN

The advent of the Industrial Revolution in Britain provided a great boost to cotton manufacture, as textiles emerged as Britain's leading export. In 1738, Lewis Paul and John Wyatt, of Birmingham, England, patented the roller spinning machine, and the flyer-and-bobbin system for drawing cotton to a more even thickness using two sets of rollers that traveled at different speeds.

Later, the invention of the spinning jenny in 1764 and Richard Arkwright's spinning frame (based on the roller spinning machine) in 1769 enabled British weavers to produce cotton yarn and cloth at much higher rates. From the late 18th century onwards, the British city of Manchester acquired the nickname *"Cottonopolis"* due to the cotton industry's omnipresence within the city, and Manchester's role as the heart of the global cotton trade.

Production capacity in Britain and the United States was further improved by the invention of the cotton gin by the American Eli Whitney in 1793. Improving technology and increasing control of world markets allowed British traders to develop a commercial chain in which raw cotton fibres were (at first) purchased from colonial plantations, processed into cotton cloth in the mills of Lancashire, and then exported on British ships to captive colonial markets in West Africa, India, and China (via Shanghai and Hong Kong). By the 1840s, India was no longer capable of supplying the vast quantities of cotton fibres needed by mechanized British factories, while shipping bulky, low-price cotton from India to Britain was time-consuming and expensive.

This, coupled with the emergence of American cotton as a superior type (due to the longer, stronger fibres of the two domesticated native American species, *Gossypium hirsutum* and *Gossypium barbadense*), encouraged British traders to purchase cotton from plantations in the United States and the Caribbean. By the mid 19th century, "King Cotton" had become the backbone of the southern American economy. In the United States, cultivating and harvesting cotton became the leading occupation of slaves. During the American Civil War, American cotton exports slumped due to a Union blockade on Southern ports, also because of a strategic decision by the Confederate government to cut exports, hoping to force Britain to recognize the Confederacy or enter the war, prompting the main purchasers of cotton, Britain and France to turn to Egyptian cotton.

British and French traders invested heavily in cotton plantations and the Egyptian government of Viceroy Isma'il took out substantial loans from European bankers and stock exchanges. After the American Civil War ended in 1865, British and French traders abandoned Egyptian cotton and returned to cheap American exports, sending Egypt into a deficit spiral that led to the country declaring bankruptcy in 1876, a key factor behind Egypt's annexation by the British Empire in 1882. During this time, cotton cultivation in the British Empire, especially India, greatly increased to replace the lost production of the American South. Through tariffs and other restrictions, the British government discouraged the production of cotton cloth in India; rather, the raw fibre was sent to England for processing.

The Indian patriot Mahatma Gandhi described the process:

- English people buy Indian cotton in the field, picked by Indian labour at seven cents a day, through an optional monopoly.
- This cotton is shipped on British ships, a three-week journey across the Indian Ocean, down the Red Sea, across the Mediterranean, through Gibraltar, across the Bay of Biscay and the Atlantic Ocean to London. One hundred per cent profit on this freight is regarded as small.
- The cotton is turned into cloth in Lancashire. You pay shilling wages instead of Indian pennies to your workers. The English worker not only has the advantage of better wages, but the steel companies of England get the profit of building the factories and machines. Wages; profits; all these are spent in England.
- The finished product is sent back to India at European shipping rates, once again on British ships. The captains, officers, sailors of these ships, whose wages must be paid, are English. The only Indians who profit are a few lascars who do the dirty work on the boats for a few cents a day.
- The cloth is finally sold back to the kings and landlords of India who got the money to buy this expensive cloth out of the poor peasants of India who worked at seven cents a day. (Fisher 1932 pp 154–156)

In the United States, Southern cotton provided capital for the continuing development of the North. The cotton produced by enslaved African Americans not only helped the South, but also enriched Northern merchants. Much of the Southern cotton was transshipped through the northern ports.

Cotton remained a key crop in the Southern economy after emancipation and the end of the Civil War in 1865. Across the South, sharecropping evolved, in which free black farmers and landless white farmers worked on white-owned cotton plantations of the wealthy in return for a share of the profits. Cotton plantations required vast labour forces to hand-pick cotton, and it was not until the 1950s that reliable harvesting machinery was introduced into the South (prior to this, cotton-harvesting machinery had been too clumsy to pick cotton without shredding the fibres). During the early 20th century, employment in the cotton industry fell, as machines began to replace labourers, and the South's rural labour force dwindled during the First and Second World Wars. Today, cotton remains a major export of the southern United States, and a majority of the world's annual cotton crop is of the long-staple American variety.

TANGUIS COTTON

In 1901, Peru's cotton industry suffered because of a fungus plague caused by a plant disease known as "cotton wilt" or, more correctly, "fusarium wilt", caused by the fungus *Fusarium vasinfectum*.

The plant disease, which spread throughout Peru, entered plant's roots and worked its way up the stem until the plant was completely dried up. Fermín Tangüis, a Puerto Rican agriculturist who lived in Peru, studied some species of the plant that were affected by the disease to a lesser extent and experimented in germination with the seeds of various cotton plants.

In 1911, after 10 years of experimenting and failures, Tangüis was able to develop a seed which produced a superior cotton plant resistant to the disease. The seeds produced a plant that had a 40 per cent longer (between 29 mm and 33 mm) and thicker fibre that did not break easily and required little water. The Tangüis cotton, as it became known, is the variety which is preferred by the Peruvian national textile industry. It constituted 75 per cent of all the Peruvian cotton production, both for domestic use and apparel exports. The Tangüis cotton crop was estimated at 225,000 bales that year.

CULTIVATION

Successful cultivation of cotton requires a long frost-free period, plenty of sunshine, and a moderate rainfall, usually from 600 to 1200 mm (24 to 48 inches). Soils usually need to be fairly heavy, although the level of nutrients does not need to be exceptional. In general, these conditions are met within the seasonally dry tropics and subtropics in the Northern and Southern hemispheres, but a large proportion of the cotton grown today is cultivated in areas with less rainfall that obtain the water from irrigation.

Production of the crop for a given year usually starts soon after harvesting the preceding autumn. Planting time in spring in the Northern hemisphere varies from the beginning of February to the beginning of June. The area of the United States known as the South Plains is the largest contiguous cotton-growing region in the world. While dryland (non-irrigated) cotton is successfully grown in this region, consistent yields are only produced with heavy reliance on irrigation water drawn from the Ogallala Aquifer.

Since cotton is somewhat salt and drought tolerant, this makes it an attractive crop for arid and semiarid regions. As water resources get tighter around the world, economies that rely on it face difficulties and conflict, as well as potential environmental problems. For example, improper cropping and irrigation practices have led to desertification in areas of Uzbekistan, where cotton is a major export. In the days of the Soviet Union, the Aral Sea was tapped for agricultural irrigation, largely of cotton, and now salination is widespread.

GENETIC MODIFICATION

Genetically modified (GM) cotton was developed to reduce the heavy reliance on pesticides. The bacterium *Bacillus thuringiensis* (Bt) naturally produces a chemical harmful only to a small fraction of insects, most notably the larvae of moths and butterflies, beetles, and flies, and harmless to other forms of life. The gene coding for BT toxin has been inserted into cotton, causing cotton to produce this natural insecticide in its tissues. In many regions, the main pests in commercial cotton are lepidopteron larvae, which are killed by the BT protein in the transgenic cotton they eat. This eliminates the need to use large amounts of broad-spectrum insecticides to kill lepidopteron pests (some of which have developed pyrethroid resistance). This spares natural insect predators in the farm ecology and further contributes to no insecticide pest management.

BT cotton is ineffective against many cotton pests, however, such as plant bugs, stink bugs, and aphids; depending on circumstances it may still be desirable to use insecticides against these. A 2006 study done by Cornell researchers, the Center for Chinese Agricultural Policy and the Chinese Academy of Science on Bt cotton farming in China found that after seven years these secondary pests that were normally controlled by pesticszide had increased, necessitating the use of pesticides at similar levels to non-Bt cotton and causing less profit for farmers because of the extra expense of GM seeds.

However a more recent 2009 study by the Chinese Academy of Sciences, Stanford University and Rutgers University refutes this. They concluded that the GM cotton effectively controlled bollworm. The secondary pests were mostly miridae (plant bugs) whose increase was related to local temperature and rainfall and only continued to increase in half the villages studied. Moreover, the increase in insecticide use for the control of these secondary insects was far smaller than the reduction in total insecticide use due to BT cotton adoption.

The International Service for the Acquisition of Agri-biotech Applications (ISAAA) said that, worldwide, GM cotton was planted on an area of 16 million hectares in 2009. This was 49 per cent of the worldwide total area planted in cotton. The U.S. cotton crop was 93 per cent GM in 2010 and the Chinese cotton crop was 68 per cent GM in 2009. The initial introduction of GM cotton proved to be a huge success in Australia - the yields were equivalent to the no transgenic varieties and the crop used much less pesticide to produce (85 per cent reduction). The subsequent introduction of a second variety of GM cotton led to increases in GM cotton production until 95 per cent of the Australian cotton crop was GM in 2009.

The initial introduction of GM cotton proved to be a huge success in Australia - the yields were equivalent to the no transgenic varieties and the crop used much less pesticide to produce (85 per cent reduction). The subsequent introduction of a second variety of GM cotton led to increases in GM cotton production until 95 per cent of the Australian cotton crop was GM in 2009.

Cotton has also been genetically modified for resistance to glyphosate (marketed as Roundup in North America), an inexpensive and highly effective, but broad-spectrum herbicide. Originally, it was only possible to achieve glyph sate resistance when the plant was young, but with the development of Roundup Ready Flex, it is possible to achieve glyphosate resistance much later in the growing season.

GM cotton acreage in India continues to grow at a rapid rate, increasing from 50,000 hectares in 2002 to 8.4 million hectares in 2009. The total cotton area in India was 9.6 million hectares (the largest in the world or, about 35 per cent of world cotton area), so GM cotton was grown on 87 per cent of the cotton area in 2009. This makes India the country with the largest area of GM cotton in the world, surpassing China (3.7 million hectares in 2009).

The major reasons for this increase is a combination of increased farm income ($225/ha) and a reduction in pesticide use to control the cotton bollworm. Cotton has gossypol, a toxin that makes it inedible. However, scientists have silenced the gene that produces the toxin, making it a potential food crop.

ORGANIC PRODUCTION

Organic cotton is generally understood as cotton, from plants not genetically modified, that is certified to be grown without the use of any synthetic agricultural chemicals, such as fertilizers or pesticides. Its production also promotes and enhances biodiversity and biological cycles. United States cotton plantations are required to enforce the National Organic Programmeme (NOP). This institution determines the allowed practices for pest control, growing, fertilizing, and handling of organic crops.As of 2007, 265,517 bales of organic cotton were produced in 24 countries, and worldwide production was growing at a rate of more than 50 per cent per year.

PESTS AND WEEDS

The cotton industry relies heavily on chemicals, such as fertilizers and insecticides, although a very small number of farmers are moving towards an organic model of production, and organic cotton products are now available for purchase at limited locations. These are popular for baby clothes and diapers. Under most definitions, organic products do not use genetic engineering.

Historically, in North America, one of the most economically destructive pests in cotton production has been the boll weevil. Due to the US Department of Agriculture's highly successful Boll Weevil Eradication Programme (BWEP), this pest has been eliminated from cotton in most of the United States. This programme, along with the introduction of genetically engineered Bt cotton (which contains a bacterial gene that codes for a plant-produced protein that is toxic to a number of pests such as cotton bollworm and pink bollworm), has allowed a reduction in the use of synthetic insecticides.

Other significant global pests of cotton include the pink bollworm, Pectinophora gossypiella; the chili thrips, Scirtothrips dorsalis; and the cotton seed bug, Oxycarenus hyalinipennis.

HARVESTING

Most cotton in the United States, Europe, and Australia is harvested mechanically, either by a cotton picker, a machine that removes the cotton from the boll without damaging the cotton plant, or by a cotton stripper, which strips the entire boll off the plant. Cotton strippers are used in regions where it is too windy to grow picker varieties of cotton, and usually after application of a chemical defoliant or the natural defoliation that occurs after a freeze. Cotton is a perennial crop in the tropics, and without defoliation or freezing, the plant will continue to grow. Cotton continues to be picked by hand in developing countries.

COMPETITION FROM SYNTHETIC FIBRES

The era of manufactured fibres began with the development of rayon in France in the 1890s. Rayon is derived from a natural cellulose and cannot be considered synthetic, but requires extensive processing in a manufacturing process, and led the less expensive replacement of more naturally derived materials. A succession of new synthetic fibres were introduced by the chemicals industry in the following decades. Acetate in fibre form was developed in 1924.

Nylon, the first fibre synthesized entirely from petrochemicals, was introduced as a sewing thread by DuPont in 1936, followed by DuPont's acrylic in 1944. Some garments were created from fabrics based on these fibres, such as women's hosiery from nylon, but it was not until the introduction of polyester into the fibre marketplace in the early 1950s that the market for cotton came under threat.The rapid uptake of polyester garments in the 1960s caused economic hardship in cotton-exporting economies, especially in Central American countries, such as Nicaragua, where cotton production had boomed tenfold between 1950 and 1965 with the advent of cheap chemical pesticides. Cotton production recovered in the 1970s, but crashed to pre-1960 levels in the early 1990s.

Beginning as a self-help programme in the mid-1960s, the Cotton Research and Promotion Programme (CRPP) was organized by U.S. cotton producers in response to cotton's steady decline in market share. At that time, producers voted to set up a per-bale assessment system to fund the programme, with built-in safeguards to protect their investments.

With the passage of the Cotton Research and Promotion Act of 1966, the programme joined forces and began battling synthetic competitors and re-establishing markets for cotton. Today, the success of this programme has made cotton the best-selling fibre in the U.S. and one of the best-selling fibres in the world.

Administered by the Cotton Board and conducted by Cotton Incorporated, the CRPP works to greatly increase the demand for and profitability of cotton through various research and promotion activities. It is funded by U.S. cotton producers and importers.

USES

Cotton is used to make a number of textile products. These include terrycloth for highly absorbent bath towels and robes; denim for blue jeans; chambray, popularly used in the manufacture of blue work shirts (from which we get the term "blue-collar"); and corduroy, seersucker, and cotton twill. Socks, underwear, and most T-shirts are made from cotton. Bed sheets often are made from cotton.

Cotton also is used to make yarn used in crochet and knitting. Fabric also can be made from recycled or recovered cotton that otherwise would be thrown away during the spinning, weaving, or cutting process. While many fabrics are made completely of cotton, some materials blend cotton with other fibres, including rayon and synthetic fibres such as polyester. It can either be used in knitted or woven fabrics, as it can be blended with elastine to make a stretchier thread for knitted fabrics, and apparel such as stretch jeans.

In addition to the textile industry, cotton is used in fishnets, coffee filters, tents, gunpowder (see nitrocellulose), cotton paper, and in bookbinding. The first Chinese paper was made of cotton fibre. Fire hoses were once made of cotton.

The cottonseed which remains after the cotton is ginned is used to produce cottonseed oil, which, after refining, can be consumed by humans like any other vegetable oil. The cottonseed meal that is left generally is fed to ruminant livestock; the gossypol remaining in the meal is toxic to monogastric animals.

Cottonseed hulls can be added to dairy cattle rations for roughage. During the American slavery period, cotton root bark was used in folk remedies as an abortifacient, that is, to induce a miscarriage.

Cotton linters are fine, silky fibres which adhere to the seeds of the cotton plant after ginning. These curly fibres typically are less than 1/8 in (3 mm) long. The term also may apply to the longer textile fibre staple lint as well as the shorter fuzzy fibres from some upland species. Linters are traditionally used in the manufacture of paper and as a raw material in the manufacture of cellulose.

In the UK, linters are referred to as "cotton wool". This can also be a refined product (*absorbent cotton* in U.S. usage) which has medical, cosmetic and many other practical uses. The first medical use of cotton wool was by Dr. Joseph Sampson Gamgee at the Queen's Hospital (later the General Hospital) in Birmingham, England.

Shiny cotton is a processed version of the fibre that can be made into cloth resembling satin for shirts and suits. However, it is hydrophobic (does not absorb water easily), which makes it unfit for use in bath and dish towels (although examples of these made from shiny cotton are seen).

The term Egyptian cotton refers to the extra long staple cotton grown in Egypt and favoured for the luxury and upmarket brands worldwide. During the U.S. Civil War, with heavy European investments, Egyptian-grown cotton became a major alternate source for British textile mills. Egyptian cotton is more durable and softer than American Pima cotton, which is why it is more expensive. Pima cotton is American cotton that is grown in the southwestern states of the U.S.

INTERNATIONAL TRADE

The largest producers of cotton, currently (2009), are China and India, with annual production of about 34 million bales and 24 million bales, respectively; most of this production is consumed by their respective textile industries. The largest exporters of raw cotton are the United States, with sales of $4.9 billion, and Africa, with sales of $2.1 billion. The total international trade is estimated to be $12 billion. Africa's share of the cotton trade has doubled since 1980.

Neither area has a significant domestic textile industry, textile manufacturing having moved to developing nations in Eastern and South Asia such as India and China. In Africa, cotton is grown by numerous small holders. Dunavant Enterprises, based in Memphis, Tennessee, is the leading cotton broker in Africa, with hundreds of purchasing agents. It operates cotton gins in Uganda, Mozambique, and Zambia. In Zambia, it often offers loans for seed and expenses to the 180,000 small farmers who grow cotton for it, as well as advice on farming methods. Cargill also purchases cotton in Africa for export.

The 25,000 cotton growers in the United States are heavily subsidized at the rate of $2 billion per year. The future of these subsidies is uncertain and has led to anticipatory expansion of cotton brokers' operations in Africa. Dunavant expanded in Africa by buying out local operations. This is only possible in former British colonies and Mozambique; former French colonies continue to maintain tight monopolies, inherited from their former colonialist masters, on cotton purchases at low fixed prices.

LEADING PRODUCER COUNTRIES

Table. Top Ten Cotton Producers — 2009(480-Pound Bales)

People's Republic of China	32.0 million bales
India	23.5 million bales
United States	12.4 million bales
Pakistan	10.8 million bales
Brazisl	5.5 million bales
Uzbekistan	4.4 million bales
Australia	1.8 million bales
Turkey	1.7 million bales
Turkmenistan	1.1 million bales
Syria	1.0 million bales

The five leading exporters of cotton in 2009 are:

1. The United States,
2. India,
3. Uzbekistan,
4. Pakistan,
5. Brazil.

The largest non-producing importers are Korea, Russia, Taiwan, Japan, and Hong Kong.

In India, the states of Maharashtra (26.63 per cent), Gujarat (17.96 per cent) and Andhra Pradesh (13.75 per cent) and also Madhya Pradesh are the leading cotton producing states, these states have a predominantly tropical wet and dry climate.

In Pakistan, cotton is grown predominantly in the provinces of Punjab and Sindh. The leading city in cotton production is the Punjabi city of Faisalabad which is also leading in textiles within Pakistan. The Punjab has a tropical wet and dry climate throughout the year therefore enhancing the growth of cotton.

In the United States, the state of Texas led in total production as of 2004, while the state of California had the highest yield per acre.

FAIR TRADE

Cotton is an enormously important commodity throughout the world. However, many farmers in developing countries receive a low price for their produce, or find it difficult to compete with developed countries.

This has led to an international dispute (see United States – Brazil cotton dispute):

On 27 September 2002, Brazil requested consultations with the US regarding prohibited and actionable subsidies provided to US producers, users and/or exporters of upland cotton, as well as legislation, regulations, statutory instruments and amendments thereto providing such subsidies (including export credits), grants, and any other assistance to the US producers, users and exporters of upland cotton. On 8 September 2004, the Panel Report recommended that the United States "withdraw" export credit guarantees and payments to domestic users and exporters, and "take appropriate steps to remove the adverse effects or withdraw" the mandatory price-contingent subsidy measures.

In addition to concerns over subsidies, the cotton industries of some countries are criticized for employing child labour and damaging workers' health by exposure to pesticides used in production. The Environmental Justice Foundation has campaigned against the prevalent use of forced child and adult labour in cotton production in Uzbekistan, the world's third largest cotton exporter.

The international production and trade situation has led to "fair trade" cotton clothing and footwear, joining a rapidly growing market for organic clothing, fair fashion or so-called "ethical fashion". The fair trade system was initiated in 2005 with producers from Cameroon, Mali and Senegal.

TRADE

Cotton is bought and sold by investors and price speculators as a tradable commodity on 2 different stock exchanges in the United States of America.

- Cotton futures contracts are traded on the New York Mercantile Exchange (NYMEX) under the ticker symbol TT. They are delivered every year in March, May, July, October, and December.
- Cotton #2 futures contracts are traded on the New York Board of Trade (NYBOT) under the ticker symbol CT. They are delivered every year in March, May, July, October, and December.

CRITICAL TEMPERATURES

- *Favourable travel temperature range*: below 25°C (77°F)
- *Optimum travel temperature*: 21°C (70°F)
- *Glow temperature*: 205°C (401°F)
- *Fire point*: 210°C (410°F)
- *Autoignition temperature*: 407°C (765°F)
- *Autoignition temperature (for oily cotton)*: 120°C (248°F)

Cotton dries out, becomes hard and brittle and loses all elasticity at temperatures above 25°C (77°F). Extended exposure to light causes similar problems. A temperature range of 25°C (77°F) to 35°C (95°F) is the optimal range for mold development. At temperatures below 0°C (32°F), rotting of wet cotton stops. Damaged cotton is sometimes stored at these temperatures to prevent further deterioration.

BRITISH STANDARD YARN MEASURES

- 1 thread = 55 inches (about 137 cm)
- 1 skein or rap = 80 threads (120 yards or about 109 m)
- 1 hank = 7 skeins (840 yards or about 768 m)
- 1 spindle = 18 hanks (15,120 yards or about 13.826 km)

FIBRE PROPERTIES

Property	Evaluation
Shape	Fairly uniform in width, 12-20 micrometers; length varies from 1 cm to 6 cm (½ to 2½ inches); typical length is 2.2 cm to 3.3 cm (7/8 to 1¼ inches).
Luster	high
Tenacity (strength)	
Dry	3.0-5.0 g/d
Wet	3.3-6.0 g/d
Resiliency	low
Density	1.54-1.56 g/cm^3
Moisture absorption	
raw: conditioned	8.5%
saturation	15-25%
mercerized: conditioned	8.5-10.3%
saturation	15-27%+
Dimensional stability	good
Resistance to	
acids	damage, weaken fibres
alkali	resistant; no harmful effects
organic solvents	high resistance to most
sunlight	Prolonged exposure weakens fibres.
microorganisms	Mildew and rot-producing bacteria
insects	damage fibres.
	Silverfish damage fibres.
Thermal reactions	
to heat	Decomposes after prolonged exposure
to flame	to temperatures of 150˚C or over. Burns readily

The chemical composition of cotton is as follows:

- Cellulose 91.00 per cent
- Water 7.85 per cent
- Protoplasm, pectins 0.55 per cent
- Waxes, fatty substances 0.40 per cent
- Mineral salts 0.20 per cent

COTTON GENOME

A public genome sequencing effort of cotton was initiated [45] in 2007 by a consortium of public researchers. They agreed on a strategy to sequence the genome of cultivated, tetraploid cotton. "Tetraploid" means that cultivated cotton actually has two separate genomes within its nucleus, referred to as the A and D genomes. The sequencing consortium first agreed to sequence the D-genome relative of cultivated cotton (G. raimondii, a wild Central American cotton species) because of its small size and limited number of repetitive elements.

It is nearly one-third the number of bases of tetraploid cotton (AD), and each chromosome is only present once. The A genome of G. arboreum would be sequenced next. Its genome is roughly twice the size of G. raimondii's. Part of the difference in size between the two genomes is the amplification of retrotransposons (GORGE).

Once both diploid genomes are assembled, then research could begin sequencing the actual genomes of cultivated cotton varieties. This strategy is out of necessity; if one were to sequence the tetraploid genome without model diploid genomes, the euchromatic DNA sequences of the AD genomes would co-assemble and the repetitive elements of AD genomes would assembly independently into A and D sequences respectively. Then there would be no way to untangle the mess of AD sequences without comparing them to their diploid counterparts.

The public sector effort continues with the goal to create a high-quality, draft genome sequence from reads generated by all sources. The public-sector effort has generated Sanger reads of BACs, fosmids, and plasmids as well as 454 reads. These later types of reads will be instrumental in assembling an initial draft of the D genome.

They announced that they would donate their raw reads to the public. This public relations effort gave them some recognition for sequencing the cotton genome. Once the D genome is assembled from all of this raw material, it will undoubtedly assist in the assembly of the AD genomes of cultivated varieties of cotton, but a lot of hard work remains.

YIELD PER CENT MAXIMIZATION IN THE COTTON SPINNING INDUSTRY (YARN MANUFACTURING)

Yield per cent shows the performance of any industry, it shows efficiency of the industry to convert the raw material into the finished goods. The industry management focuses so much on the maximization of yield so that maximum profit can be earned. Yield per cent directly relates with profit of the industry, textile cotton spinning industry usually give the yield per cent up to 84, this can be increased up to 1 to 2 per cent , adding million in the profit of cotton spinning industry. This article presents the methodology to enhance the yield per cent of cotton spinning industry.

INTRODUCTION

Raw material consist impurities up to 4-16 per cent (according to type, region and the efficiency of the ginning), the yarn manufacturing processes (Blow-room to carding) are especially designed to remove these impurities. While removing the impurities some good fibres are also removed due to inefficient machine setting or improper machine sequence selection and there are many more causes which decrease the yield per cent . Some of the causes are discussed here which increase/decrease the yield per cent such as

- Raw material selection
- Raw martial process route selection
- Improper machine setting
- Maintenance of Machines
- Inefficient air conditioning plant
- Improper material handling

RAW MATERIAL SELECTION

Raw martial costs about 50-70 per cent of the total cost of the product[1]. It is mentioned above that impurities in the cotton raw material in Pakistan ranges between 4 to 16 per cent, these impurities must be removed to achieve the required quality product. If the raw material for the yarn is not properly selected than it may cause decrease in the yield per cent .

Finer yarn requires more cleaning of the material than the coarser yarn because impurities present in the fine yarn create hinders in further processing and can be more visible and vice versa than coarse yarn. Finer yarn required fine quality raw cotton which have sufficient strength and the fibre length if the cotton selected for the finer yarn is of not sufficient length and strength then the good fibres may be wasted during the processing in result of that yield per cent is decreased. Another main characteristic of raw material which also contribute in decrease of yield per cent is extra moisture in material because if the material has extra moister present in it then it may difficulty to separate the impurities and good fibres which required more efforts, energy, time etc such material sticks with machine parts which may go into the waste and become the cause in yield per cent decrease.

YUCCA FIBRE

Yucca fibres were at one time widely used throughout Central America for many things. Currently they are mainly used to make twine. Yucca leaves are harvested and then cut to a standard size. The leaves are crushed in between two large rollers producing the fibres which are bundled up and dried in the sun over trellises. The dried fibres are combined into rolags. At this point it is ready to spin. The waste, a pulpy liquid that stinks, can be used as a fertilizer.

RAW MATERIAL PROCESS ROUTE SELECTION

Route selection for processing the raw material is also another important task for the management. In cotton spinning industry different lines of processing are installed in the Blow Room, routes have different beating points and according to amount of impurities present in material and quality requirement the processing route is selected. Which reduces the wastage of good fibre, resulting increase in (production) yield per cent . Processing route selection is based on the type of the material process *e.g.* natural and synthetic material cannot be processed on the same route even you are required to manufacturing the pc (Polyester cotton with any ration) for blend yarn you have to process both fibres separately and at any particular point blend them with required ratio. Now a days technology facilities process and separate blending machines are available to blend the material with any ratio or you can blend the slivers of the different materials at the draw frame.

Material passage at the draw frame is also pre-decided whether material should be double (Breaker and finisher) or triple passage (breaker, intermediate and finisher). With less passage the production increased by saving the waste per cent *i.e.* 0.5 but the quality may be effected or not if the raw material quality is enough then the intermediate process may be omitted otherwise It should be carried to improve the quality so that final product can be sold. If after the processing product is not sold then it means whole material and efforts consumed on it are wasted and yield per cent becomes 0. Therefore it should be kept in mind that product must be sell out.

IMPROPER MACHINE SETTING

Required quality of the product cannot be achieved without the proper setting of machine, any spinning machine setting parameters includes

- Speed
- Gauge
- Top Roller Pressure
- Air pressure

With the above mentioned setting parameters you can manufacture the required quality of yarn within the required time frame. With the speed you can adjust the draft, production rate, beats per unit time, and many more. With the gauge and pressure (Air and top Roller) setting you can easily process the material without deteriorating its quality and properly removing the impurities to achieve the targeted product and the result of this surely contributes in the increase of yield per cent . If any above said setting parameter is improperly set then it will deteriorate the quality and ultimately decrease the yield per cent .

MAINTENANCE OF MACHINES:

There is old saying that machine cannot become older, if properly maintained time to time, if machine is not properly maintained then it will not deliver the required quality and may also waste good fibres and due to this yield per cent will decrease. *E.g.* if card machine is not properly overhauled and its taker-in and cylinder wires are out of order, when the material is process on such a machine then the machine is unable to remove the impurities and also unable to properly open the fibres up to individual fibre stage, then the product (sliver) delivered by the card machine is not match with requirement thus for the required quality most of good material may become useless or you have option to reprocess the material in both cases yield per cent is ultimately decrease and any owner/management cants bear this, therefore for improving the yield per cent machine maintenance should be primary objective well maintained machines produces required/targeted quality and yield.

INEFFICIENT AIR CONDITIONING PLANT

Air-conditioning plant contributes most of its share in increase/decrease of the yield per cent , due to A.C plant the yield per cent can be at maximum or minimum level therefore if plant is properly maintained providing proper suctions into transportation pipes, machines, atmospheric conditions with the requirement of each department then surely yield per cent can be increase.

The main objectives of the A.C plant is to maintenance required temperature and RH per cent into every department, provide proper suction into return ducts so that the fluf and good fibres from the floor properly collected and have sufficient air filters, so that the dust, other impurities and good fibres can be separated. Each machine have also separate suction system *e.g.* at ring frame pnewmafil is collected by the internal machine suction. Pnemafil 100 per cent good fibres which and be reprocessed with minimum efforts if these are not collected then went into waste hence causing decrease in yield per cent . Similarly A.C plant performs its job in every machine to prevent the wastage of good/processable fibre.

THE OBJECTIVES OF THE AIR-CONDITIONING PLANTS INCLUDES

- Control the RH per cent and Temperature in every department
- Collect and Control the fulf
- Properly separate the collected impurities and good fibre by the help of filters

By achieving the above mentioned objectives we can improve the efficiency of man and machine, Quality of yarn and production of plant in result of that we can achieve our target *i.e.* yield per cent maximization

IMPROPER MATERIAL HANDLING:

Improper handling of material before, during and after process decreases the yield per cent . *E.g.* the slivers cans during transporting from the card to draw frame or from draw to comber or simplex if not properly carried and by chance material may be deteriorate by hand touching or due to other reason can fall on floor because of uneven floor or cane wheels will contribute in yield per cent shortage. Handling of material start when the raw material arrived and you have to plane well where to store it and in which conditions it should me stored considering that when transporting to production floor may not be effected/damage/detoriated. Material handling is also required in each department of cotton spinning staring from blow-room as the material is process through each department *e.g.* final product of blow-room is lap or final product of chute feed system is sliver of card therefore the lap/slivers are properly transported and stored so that their quality cannot be effected similarly the final product of each department is well stored, material handling also deals with the proper storage and transportation of the finished product.

PHOTOCHROMIC

Photochromism is the reversible transformation of a chemical species between two forms by the absorption of electromagnetic radiation, where the two forms have different absorption spectra. This can be described as a reversible change of colour upon exposure to light? This property is a boon for scientists doing research on intelligent textiles where they are making use of this property to store data on the surface of textile fabrics and polymer sheets. Whereas the same property of some reactive dyes is a bane for textile processors.

The change in shade after dyeing creates unwanted problems in dyeing. An optical recording medium contains, on a base, one or more dyes and a polymer which forms liquid-crystalline phases. The information is written into the uniformly oriented liquid-crystalline polymer layer, for example by means of a laser. During this procedure, the polymer heats up locally to above a phase transition temperature. By cooling, the resulting change is frozen in the glass state. The information can be erased by applying an electric field and/or heating. The recording material permits high-contrast storage and possesses high sensitivity, good resolution and excellent stability. There are other chromatic properties called electrochromatism and thermochromatism of dyes that are affected by electric field and heat respectively. Photochromic colours are plastisol-based inks, which are off-white when not exposed to UV radiation. It gains colour when exposed to Sun light/UV light. The colour change is “reversible,” *i.e.*, the colour will fade again and appear colour less upon removal from UV light/sun light exposure. These inks are available in various colours. See our colour availability chart for a complete list of available colours.

The uses of Americos Photo chromic colours are:

- On garment to create novel products and promotional items like T-shirts
- On fabric/garment to print company logo/brand name to prevent duplication
- On garment which are used for party wear
- Thermometers and temperature indicators
- Security printing
- Food industry to indicate temperature of packaged food

NANO TEXTILE

WHAT IS NANOTECHNOLOGY?

Nanotechnology is the technical process of working on the nano-scale – each nano-scale molecule is one million times smaller than a grain of sand. Nanotechnology refers to not only the small size of the materials being used, but also how those materials are engineered to perform specific functions. Traditional coatings make garments feel stiff and clog the weave of the fabric preventing breath ability. Using nanotechnology, our treatments are small enough to attach to individual fibres, delivering superior performance characteristics without compromising the look, feel or comfort of the fabric.

COOLEST COMFORT

Night before the big job interview. Sleep elusive. Hot, then cold, then hot again. No fever, just frantic. Good thing your sheets are keeping you cool. Wake up to sunshine, fresh outlook. Feel like a million bucks. Dress to impress. Blow them away with energy and intelligence. Career happily underway. With Nano Textiles Coolest Comfort, you stay dry and comfortable.

An advanced moisture-wicking system helps balance your body temperature to keep you feeling fresh.

Keep dry, stay cool, rest easy.

- Balances body temperature
- Enhances comfort
- Retains fabric's natural softness
- Allows fabric to breathe naturally

RESISTS SPILLS

Playing hooky. Breakfast in bed. Perfect white linens. Reading the *Times*. Everything just right. Phone rings. Newspaper falls on breakfast tray. Cranberry juice flies on favourite duvet. Relaxed mood is ruined. Good thing your bed cover isn't.

Conventional methods only coat fabrics superficially. But Nano Textiles uses an innovative process to build spill resistance into the individual fibres, so fabrics hold up under the toughest spills. You'll never sacrifice softness or durability, though - Nano Textiles technology makes your duvets, drapery, tablecloths, placemats, and even mattress pads stay looking beautiful, longer.

If it could, your bed would thank you.

- Repels liquids
- Outperforms conventional fabric treatments
- Provides long lasting protection
- Extends the life of the fabric
- Retains fabric's natural softness
- Allows fabric to breathe naturally

REPELS AND RELEASES STAINS

10th annual family reunion. More relatives than you can count. Potluck dinner. Plates piled high. Favourite tablecloth "decorated" with Grandma's special spaghetti sauce, and who knows what else. Luckily, your tablecloth is enhanced with an invisible shield. Nano Textiles Repels and Releases Stains technology actually prevents stains from sinking into fabrics, so spills are never a problem. Easy cleaning and durable protection keeps your finery looking its finest.

- Repels spills
- Helps stains wash out easily
- Provides long lasting protection
- Extends the life of the fabric
- Retains fabric's natural softness
- Allows fabric to breathe naturally

RESISTS STATIC

Cozy evening in. An old movie in front of a roaring fire. Hot popcorn. Great date. Warm blanket- It's not the TV with the bad static, it's the blanket. Dust, lint, and dog hair getting in the way of real electricity. Fast forward. New blanket. Same great date. The right kind of attraction.

Nano Textiles provides permanent static resistance for throws and blankets. It repels hair, lint, and dust while staying soft, comfortable, and beautifully durable. Let the sparks fly, just not between you and your blankets.

- Provides permanent static protection
- Repels lint, dust, dirt and pet hair
- Enhances appearance and comfort
- Retains fabric's natural softness
- Allows fabric to breathe naturally

ORGANIC CLOTHING: WEAR WITH AWARENESS

When picking cotton clothing, today's ecologically minded consumer has several choices. Once believed to be a pure, natural fabric, today's cotton is dosed with harmful chemicals that pollute our environment. Today, thanks to strong media awareness and a revitalization of old techniques, we have alternatives. Organically grown cotton, naturally coloured cotton and recycled cotton products give us three choices for truly natural clothing. It may cost a little more, but by supporting the natural cotton industry's growth, we're supporting ourselves and the future of our planet.

PROBLEMS WITH CONVENTIONAL COTTON

The use of cotton dates back to the Egyptians some 4,500 year ago. For thousands of years, cotton, grown using natural methods, was the primary source of textiles. Today, cotton production is far from natural. The United States alone dumps 8.5 million tons of pesticides on cotton fields annually. And that's not all. Conventionally grown cotton accounts for nearly 25 per cent of the total insecticide use for crops worldwide. This chemical onslaught harms our environment by polluting ground waters and soil, resulting in the death of wildlife and natural habitats. Most of the cotton garments sold today is made from cotton grown with chemical pesticides, bleached and then coloured with chemical dyes containing toxic heavy metals. About a third of a pound of chemicals are used to make one adult T-shirt; two-thirds of a pound of chemicals can go into a pair of jeans.

THREE ECO-FRIENDLY ALTERNATIVES

Concerned consumers are choosing clothing made from organic cotton, naturally coloured cotton, and recycled sources. Here are details on these three environmentally healthy ways of producing cotton clothing and other textiles.

ORGANICALLY GROWN COTTON

Organic cotton is grown without the use of synthetic chemical fertilizers, pesticides or defoliants. By incorporating farming practices that increase fertility and diverse eco-systems, organic farmers rely on time-honoured techniques. Crop rotation, cover crops, organic fertilizers, integrated pest management, and human labour for weed control are a few of the methods used. To be certified organic, the soil must be free of synthetic pesticides for at least three years. Farmers and processors are required to pass yearly inspections.

With the obvious environmental and health benefits of growing cotton organically, why aren't more farmers doing it? In the past, demand for organic cotton has been limited due to higher production costs. However, in recent years media attention has been strong. Some major industry players such as Levi Strauss, Nike and The Gap are blending organic cotton fibres with conventionally produced cotton. The growing public awareness of the toxicity of conventional farming methods has resulted in an increase in both the demand for organically grown cotton and the amount of acreage planted.

NATURALLY COLOURED COTTON

Until recently, chemical dyeing was counted as the only viable way to colour cotton clothing. This process requires several steps, each of which creates toxic waste. Cotton is often bleached before it is dyed and heavy metal mordants are used to adhere the dye to the fabric. Because dyes have a hard time adhering to cotton, at least half of the chemicals end up as waste water in rivers and in the soil. Even in small amounts these heavy metals are lethal.

In 1982, entomologist Sally Fox reintroduced naturally coloured cotton, eliminating the need for dyeing altogether. Cottons of different colours have always existed in nature. Like our eyes or hair, cotton is genetically encoded with colours ranging from brown to tan. Native peoples have used these wild cottons for weaving and hand spinning for centuries. Because of its short fibres and inherent weakness, however, it was unable to be processed by modern textile machinery and had limited commercial value.

With a programme of plant breeding, Fox developed a strong, long-fibre, coloured cotton that can be used commercially. Today, coloured cotton is grown on the stem in shades of brown, reddish brown, green and yellow, totally eliminating the need for dyes.

In the long run, no dyeing means a savings in our pocket. The cost of one pound of dyed cotton including dyestuff, water, energy costs and toxic waste disposal is 20 to 40 per cent higher than that of coloured cotton. And with coloured cotton there are no hidden costs to the environment.

In addition coloured cotton offers:

- An eco-friendly solution. Since there is no bleaching or harmful dyestuff involved in manufacturing, no waste water is produced.
- Unlike dyed material, the colour of the fabrics made from coloured cotton actually deepens with washing.
- Coloured cotton is naturally pest and disease tolerant, making it easier to grow organically.
- Coloured cotton is suitable for chemically sensitive people.
- Coloured cotton provides new yarn and fabric design potentials.
- Without the need for abundant water or energy sources, new mill sites have many more choices for location.

In recent years, the demand for coloured cotton has increased. Large companies like Esprit and Levi have launched popular "green lines" of cotton outerwear using dye-free, unbleached, organic green cotton. The demand for these items far exceed the supply. Coloured cotton is now being grown in the United States, Europe and Australia.

RECYCLED COTTON

Another ecological choice when purchasing cotton clothing is Eco Fibre. Eco Fibre is a recycled cotton fabric made from recovered cotton that would otherwise be cast off during the spinning, weaving or cutting process. There are no harsh chemicals used in the processing of this fabric.

The next time you are shopping for cotton clothing, choose garments made from organic cotton, naturally coloured cotton or recycled cotton. By buying from these natural sources, you are supporting the demand for a new earth-friendly textile industry. Your choice makes a difference.

Hemp is naturally one of the most ecologically friendly fabrics and also the oldest. The Columbia History of the World states that the oldest relics of human industry are bits of hemp fabric discovered in tombs dating back to approximately 8,000 BC.

Hemp fibre is one of the strongest and most durable natural textile fibres. Not only is it strong, but it also holds its shape having one of the lowest per cent elongation of any natural fibre.

In fact, its combination of ruggedness and comfort were utilized by Levi Strauss as a lightweight duck canvas for the very first pair of jeans made in California. Furthermore hemp has the best ratio of heat capacity of all fibres giving it superior insulation properties.

As a fabric, hemp provides all the warmth and softness of other natural textiles but with a superior durability seldom found in other materials. Natural organic hemp fibre 'breathes' and is biodegradable. Hemp blended with other fibres easily incorporate the desirable qualities of both textiles. When combined with the natural strength of hemp, the soft elasticity of cotton or the smooth texture of silk create a whole new genre of fashion design.

A fibre of a hundred uses besides fabrics, hemp is also used in the production of paper. The oldest piece of paper - over 2000 years old - was discovered in China and is made from hemp. Until 1883, between 75 per cent and 90 per cent of all paper in the world was made with hemp fibre. The Gutenberg bible (15th century), Lewis Carroll's Alice in Wonderland (19th century) and just about everything in between was printed on hemp paper.

Thomas Jefferson wrote the early drafts of the Declaration of Independence on hemp paper produced in Holland. Jefferson grew hemp on his plantation as an industrial crop, selling the dried stalk to the U.S. Navy as outfitting material. George Washington also grew hemp, harvesting the fibrous seed for a variety of commercial uses including a skin lotion.

Other uses include feed for animals and for humans in veggie burgers, salad dressings, and pastas. Hemp seed is nutritious and contains more essential fatty acids than any other source, is second only to soybeans in complete protein (but is more digestible by humans), is high in B-vitamins, and is a good source of dietary fibre. Cosmetics manufacturers include hemp oil in makeup, skin lotions, and shampoo. In Europe, hemp is used in household cleaners as a natural alternative to harsher chemicals.

Hemp is a renewable resource which grows more quickly and easily than trees making hemp more cost effective than waiting decades for trees to grow to be used in man-made fibre production such as lyocell and rayon from wood pulps. The bark of the hemp stalk contains bast fibres, which are among the Earth's longest natural soft fibres and are also rich in cellulose.

The cellulose and hemi-cellulose in its inner woody core are called hurds. Hemp fibre is longer, stronger, more absorbent and more insulative than cotton fibre.

Hemp produces more pulp per acre than timber on a sustainable basis, and can be used for every quality of paper. Hemp paper manufacturing can reduce wastewater contamination. Hemp's low lignin content reduces the need for acids used in pulping, and its creamy colour lends itself to environmentally-friendly bleaching instead of harsh chlorine compounds. Less bleaching results in less dioxin and fewer chemical by-products. Hemp fibre paper resists decomposition, and does not yellow with age when an acid-free process is used. Hemp paper more than 1,500 years old has been found. Hemp paper can also be recycled more times than wood-based paper.

According to the Department of Energy, hemp is an excellent biomass fuel producer and the hydrocarbons in hemp can be processed into a wide range of biomass energy sources, from fuel pellets to liquid fuels and gas. Development of bio-fuels could significantly reduce our consumption of fossil fuels and nuclear power. Hemp can be grown organically easily and hemp is most often grown without herbicides, fungicides or pesticides. Hemp is also a natural weed suppressor due to the fast growth of the plant's canopy. Eco-friendly hemp can replace most toxic petrochemical products. Research is being done to use hemp in manufacturing biodegradable plastic products: plant-based cellophane, recycled plastic mixed with hemp for injection-molded products, and resins made from the oil are just a few examples.

- Properties of Hemp
- Uses for Hemp
- Hemp Fibres and Fabrics
- Manufacturing Hemp Fabric
- Dyeing and Finishing
- Ecology of Hemp
- Industrial Hemp for renewable energy
- Ramie Fibre - Another Natural fibre:

WELLNESS FINISH WITH VITAMIN E

Functional aspects for clothing textiles, which promote the wearing comfort and well-being have become a striking sales argument. The following article introduces a product idea for "Wellness Textiles". It describes a transfer system consisting of textile/cyclodextrin-vitamin E complex. Wellness is by long more than just a buzzword for wellbeing, vitality and fitness. It has become a social phenomenon which confers the wish of 'eternal youth' to ageing persons. The term 'wellness' has become a new life philosophy and from the commercial aspect is an important pulse generator and growth motor for many trade brances. The 'wellness' fever has also reached the field of textile finishing and blew in some 'fresh wind' to this field with innovative product ideas. The following article describes a so-called transfer system by which vitamin E is converted from the textile onto the human skin. For the first time the transfer of vitamin E from the textile material into a skin model could be proven in an experiment.

VITAMIN E (A-TOCOPHEROL)

Vitamin E belongs to the group of lipid-soluble vitamins and is found in nature in many vegetable oils. The chemical term for vitamin E is "a-Tocopherol".(alpha-Tocopherol) In the cosmetic industry vitamin E is used as antioxidant and active substance among others because of its moisture binding capacity in aliphatic cosmetic creams, lotions, emulsions, body and face oils for dry skin care as well as for decorative cosmetics like lipsticks. Vitamin E is also successfully applied for various skin diseases.

Very much importance is attached to vitamin E in the food industry, too. Vitamin E is an important lipid-soluble antioxidant which has many positive physiological properties besides its vitamin character. The term 'antioxidant' describes the capability of molecules neutralizing so called radicals. Thus antioxidants are often called scavengers. Radicals are atoms or molecules which have an unpaired electron in their outer shell. Free radicals also emerge by the normal cell breathing as side products and try to snatch away an electron from other structures for means of completing their outer shell. In this way for example the cell membrane can be damaged. Antioxidants and thus also vitamin E 'deactivate' the free radicals by giving off an electron and in this way protect the cells from 'oxidative stress'.

CYCLODEXTRINS

Cyclodextrins are circular molecules produced by the enzymatic decomposition of starch and they belong to the group of oligosaccharides and consist of 6 to 8 glucose units. The interesting point about cyclodextrins is their cylindrical structure and the resulting properties and application possibilities.

The polar OH groups of the individual glucose units are on the outside of the cylinder due to their steric arrangement. The outside is hydrophilic, whereas the inside of the cylinder is non-polar and thus hydrophobic (resp. lipophilic). The cavities of the cyclodextrins (host) can take in 'guest molecules' and release them again. The chemist calls this phenomenon host-guest-chemistry. Guests are all those molecules which could fit into the cavity and are non-polar enough to interact with the lipophilic cavity surface. By this complexation the properties of the locked in molecules change. For example an increase of the water solubility of non-polar organic compounds or decrease of sightly volatile substances are obtained through complexation with cyclodextrins, but also an increase of the stability of the locked in substances to light, oxygen and heat. Altogether the complexation through cyclodextrins has been researched only to a small extent, although cyclodextrins have been marketed for some time now for means of odour absorption (antismell finish). Knowledge about which substances can actually be complexed, is still very fragmentary and quite some more fundamental research will be necessary in this field.

CYCLODEXTRIN-VITAMIN-E COMPLEX

Vitamin E being a lipid soluble substance is virtually predestined for complexation with cyclodextrines. Y-cyclodextrin (8 glucose units) has proven to be a particularly suitable 'active substance'. Examinations have shown that a 2:1 complex consisting of Y-cyclodextrin and vitamin E offers many more advantages in terms of stability than a 1:1 complex. The antioxidative potential of vitamin E is based on the reactivity of the chromanoxyl radical. The main task of the cyclodextrine is to surround the ring system of the vitamin E molecule (benzopyran-6-ol) and continue to stabilize it. In this stabilized 'packaging', vitamin E is excellently suitable for Wellness finishes of textiles worn close to the body and these products can be purchased as 'NouWell E' from CHT R Beitlich GmbH.

FIXATION AND PROOF OF CD VITAMIN E COMPLEX ON TEXTILE SURFACES

As the CD vitamin E complex does not show any substantivity, only those processes are to be considered where the product application can be controlled like for example by padding, spraying, coating or printing.

For permanent fixation of the complex, reactive polyurethanes have proven to be advantageous for the following reasons:

- Free from formaldehyde
- Soft handle
- Good permanence
- Applicable on all types of fibres.

Fixation of the complex is physical and particularly on CEL and WO chemical (reaction with -OH, -NH2-groups). Fixation of the charged cyclodextrins with reactant crosslinking agents is also be possible, but then only for application on CEL fibres. The qualitative proof of vitamin E on the textile can be done through a dyeing reaction at which the reductive properties of vitamin E are taken advantage of.

- Dripping on a FeCl3-solution onto the finished textile. In the presence of vitamin E the Fe^{3+}-Ion is reduced to Fe^{2+}.
- Dripping on a dipyridyl solution. Dipyridyl forms with Fe^{2+}ions a red chelate complex.

By means of this Redox reaction vitamin E can be easily and safely seen on the fabric with the restriction that this of course can only be done on white or pastel shaded fabric

2

Fibre Optics

ARRAYED WAVEGUIDE GRATING

Arrayed waveguide gratings (AWG) are commonly used as optical (de)multiplexers in wavelength division multiplexed (WDM) systems. These devices are capable of multiplexing a large number of wavelengths into a single optical fibre, thereby increasing the transmission capacity of optical networks considerably.

The devices are based on a fundamental principle of optics that light waves of different wavelengths interfere linearly with each other. This means that, if each channel in an optical communication network makes use of light of a slightly different wavelength, then the light from a large number of these channels can be carried by a single optical fibre with negligible crosstalk between the channels. The AWGs are used to multiplex channels of several wavelengths onto a single optical fibre at the transmission end and are also used as demultiplexers to retrieve individual channels of different wavelengths at the receiving end of an optical communication network.

OPERATION OF AWG DEVICES

The incoming light (1) traverses a free space (2) and enters a bundle of optical fibres or channel waveguides (3). The fibres have different length and thus apply a different phase shift at the exit of the fibres. The light then traverses another free space (4) and interferes at the entries of the output waveguides (5) in such a way that each output channel receives only light of a certain wavelength. The orange lines only illustrate the light path. The light path from (1) to (5) is a demultiplexer, from (5) to (1) a multiplexer.

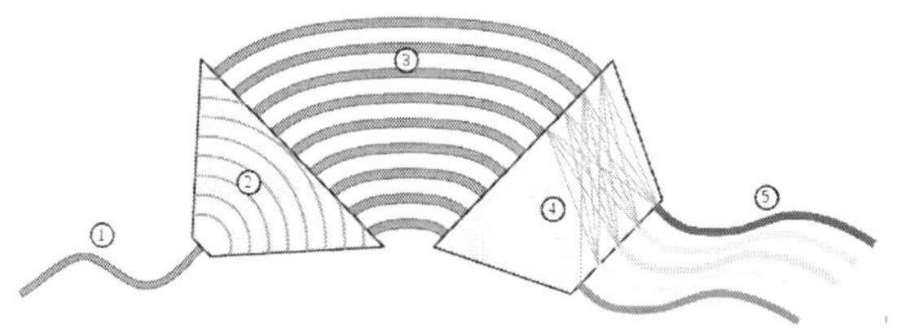

Conventional silica-based AWGs schematically shown in the above figure, are planar lightwave circuits fabricated by depositing doped and undoped layers of silica on a silicon substrate. The AWGs consist of a number of input *(1)* / output *(5)* couplers, a free space propagation region *(2)* and *(4)* and the grating waveguides *(3)*. The grating consists of a large number of waveguides with a constant length increment (ÄL). Light is coupled into the device via an optical fibre *(1)* connected to the input port. Light diffracting out of the input waveguide at the coupler/slab interface propagates through the free-space region *(2)* and illuminates the grating with a Gaussian distribution. Each wavelength of light coupled to the grating waveguides *(3)* undergoes a constant change of phase attributed to the constant length increment in grating waveguides. Light diffracted from each waveguide of the grating interferes constructively and gets refocused at the output waveguides *(5)*, with the spatial position, the output channels, being wavelength dependent on the array phase shift.

OPTICAL ATTENUATOR

An optical attenuator is a device used to reduce the power level of an optical signal, either in free space or in an optical fibre. The basic types of optical attenuators are fixed, step-wise variable, and continuously variable.

FIBRE-OPTIC ATTENUATOR APPLICATIONS

Attenuators are commonly used in fibre optic communications, either to test power level margins by temporarily adding a calibrated amount of signal loss, or installed permanently to properly match transmitter and receiver levels. Sharp bends stress optic fibres and can cause losses. If a received signal is too strong a temporary fix is to wrap the cable around a pencil until the desired level of attenuation is achieved. However, such arrangements are unreliable, since the stressed fibre tends to break over time.

FIXED FIBRE-OPTIC ATTENUATORS

Fixed optical attenuators used in fibre optic systems may use a variety of principles for their functioning. Preferred attenuators use either doped fibres, or mis-aligned splices,or total power since both of these are reliable and inexpensive. *Inline* style attenuators are incorporated into patch cables. The alternative *build out* style attenuator is a small male-female adapter that can be added onto other cables.

Non-preferred attenuators often use gap loss or reflective principles. Such devices can be sensitive to: modal distribution, wavelength, contamination, vibration, temperature, damage due to power bursts, may cause back reflections, may cause signal dispersion etc.

BUILT-IN VARIABLE FIBRE-OPTIC ATTENUATORS

Built-in variable optical attenuators may be either manually or electrically controlled. A manual device is useful for one-time set up of a system, and is a near-equivalent to a fixed attenuator, and may be referred to as an "adjustable attenuator". In contrast, an electrically controlled attenuator can provide adaptive power optimization. Attributes of merit for electrically controlled devices, include speed of response and avoiding degradation of the transmitted signal. Dynamic range is usually quite restricted, and power feedback may mean that long term stability is a relatively minor issue. Speed of response is a particularly major issue in dynamically reconfigurable systems, where a delay of one millionth of a second can result in the loss of large amounts of transmitted data. Typical technologies employed for high speed response include LCD, or Lithium niobate devices. There is a class of built-in attenuators that is technically indistinguishable from test attenuators, except they are packaged for rack mounting, and have no test display.

VARIABLE FIBRE-OPTIC TEST ATTENUATORS

Variable Fibre Optic Test Attenuators generally use a variable neutral density filter. Despite relatively high cost, this arrangement has the advantages of being stable, wavelength insensitive, mode insensitive, and offering a large dynamic range. Other schemes such as LCD, variable air gap etc. have been tried over the years, but with limited success.

They may be either manually or motor controlled. Motor control give regular users a distinct productivity advantage, since commonly used test sequences can be run automatically. Attenuator instrument calibration is a major issue. The user typically would like an absolute port to port calibration. Also, calibration should usually be at a number of wavelengths and power levels, since the device is not always linear. However a number of instruments do not in fact offer these basic features, presumably in an attempt to reduce cost. The most accurate variable attenuator instruments have thousands of calibration points, resulting in excellent overall accuracy in use.

TEST AUTOMATION

Test sequences that use variable attenuators, can be very time consuming. Therefore, automation is likely to achieve useful benefits. Both bench and handheld style devices are available that offer such features.

BEND RADIUS

Bend radius, which is measured to the inside curvature, is the minimum radius one can bend a pipe, tube, sheet, cable or hose without kinking it, damaging it, or shortening its life. The *smaller* the bend radius, the *greater* is the material flexibility (as the radius of curvature *decreases*, the curvature *increases*). The diagram below illustrates a cable with a seven-centimeter bend radius. The *minimum bend radius* is the radius below which an object such as a cable should not be bent.

FIBRE OPTICS

The minimum bend radius is of particular importance in the handling of fibre-optic cables, which are often used in telecommunications. The minimum bending radius will vary with different cable designs. The manufacturer should specify the minimum radius to which the cable may safely be bent during installation, and for the long term. The former is somewhat larger than the latter. The minimum bend radius is in general also a function of tensile stresses, *e.g.,* during installation, while being bent around a sheave while the fibre or cable is under tension. If no minimum bend radius is specified, one is usually safe in assuming a minimum long-term low-stress radius not less than 15 times the cable diameter.

Beside mechanical destruction, another reason why one should avoid excessive bending of fibre-optic cables is to minimize microbending and macrobending losses. Microbending causes light attenuation induced by deformation of the fibre while macrobending causes the leakage of light through the fibre cladding and this is more likely to happen where the fibre is excessively bent.

OTHER APPLICATIONS

Strain gauges also have a minimum bending radius. This radius is the radius below which the strain gauge will malfunction.

BUFFER (OPTICAL FIBRE)

In a fibre optic cable, a buffer is one type of component used to encapsulate one or more optical fibers for the purpose of providing such functions as mechanical isolation, protection from physical damage and fibre identification. The buffer may take the form of a miniature conduit, contained within the cable and called a "loose buffer", or "loose buffer tube". A loose buffer may contain more than one fibre, and sometimes contains a lubricating gel. A "tight buffer" consists of a polymer coating in intimate contact with the primary coating applied to the fibre during manufacture. Buffer application methods include spraying, dipping, extrusion and electrostatic methods. Materials used to create buffers can include fluoropolymers such as polyvinylidene fluoride (Kynar), polytetrafluoroethylene (Teflon), or polyurethane.

CABLE BLOWING MACHINE

Cable blowing machine (fibre blowing machine) is machine designed to fit fibre optic cables into telecommunication ducts and microducts with the use of compressed air or water.

A cable blowing machine (fibre blowing machine) consists of the following components:

- A head that ensures secure fitting of the duct into which the cable will be blown and supplying the medium (air or water) to the duct.

The cable inserted into the duct via the head goes through a sealing system preventing the medium from getting into the machine;
- Belt feeder that moves the cable towards the head;
- Base plate or a frame onto which the blowing machine's subsystems are mounted;
- Cable guidance, that is a system of bushes or rolls guiding the cable towards the blowing head;
- Air or water connections that supply the medium securely to the blowing machine;
- Meter counter that shows the length (sometimes also the velocity) of the blown cable;
- Control unit that allows to control the speed of cable insertion and the propelling power of the feeder.

TYPES OF BLOWING MACHINES

Blowing machines are classified with regard to:
- Their purpose:
  - Cable blowing machines applicable for cables of external diameter above 10 mm
  - Micro cable blowing machines for micro cables of external diameters below 10 mm
- Their drive systems:
  - With a track feeder
  - With a roller feeder
  - With a belt feeder
  - Blowing heads without feeders

CABLE JETTING

Cable jetting is a technique to install cables in ducts. It is commonly used to install cables with optical fibers in underground polyethylene ducts and is an alternative to *pulling*.

PULLING

Traditionally fibre optic cables were pulled through cable ducts in the same way as other cables, via a winch line. Every time a bend or undulation in the duct is passed the pulling force is multiplied by a friction dependent factor (which can be reduced by using lubricant). This means that the higher the local pulling force is, the higher the friction will be which the cable is experiencing while being pulled against the internal duct wall. This "capstan effect" leads to an exponential force build-up with pull distance, producing generally high pulling forces.

JETTING

Cable jetting is the process of blowing a cable through a duct while simultaneously pushing the cable into the duct. Compressed air is injected at the duct inlet and flows through the duct and along the cable at high speed. (Preferably, no suction pig is used at the cable head.) The high speed air propels the cable due to drag forces and pressure drop. The friction of the cable against the duct is reduced by the distributed airflow, and large forces that would generate high friction are avoided. Because of the expanding airflow, the air propelling forces are relatively small at the cable inlet and large at the air exhaust end of the duct. To compensate for this, an additional pushing force is applied to the cable by the jetting equipment. The pushing force, acting mainly near the cable inlet, combined with the airflow propelling forces, increasing the maximum jetting distance considerably. Special lubricants have been developed for cable jetting to further reduce friction.

ADVANTAGES OF JETTING COMPARED TO PULLING

1. Longer installation distances can be reached
2. Installation distance less dependent on bends and undulations in duct
3. Forces exerted on the cable are lower
4. Easier use jet in tandem operation
5. The step of installing a winch rope is avoided
6. Equipment is needed only at one end of the duct route

PRACTICE

In the 21st century the cable jetting technique is used worldwide, from small optical telecom cables (1.8 mm diameter) in small microducts (3 mm internal diameter) up to large copper telecom cables (35 mm diameter) in large ducts (50 mm internal diameter). Jetting is done with a pressure of the compressed air in the order of 10 bar. With the jetting technique distances per blow of 3.5 km have been reached, while spliceless links of 12 km have been reached by placing jetting equipment in tandem. It is possible to install 12 km in one day with one small crew. In mid 1990s the technique was also developed to install multiple smaller microducts, bundles, into a larger duct in one installation. This is called *multi-ducting*, *microduct cabling*, or *bundle blowing*. Each can hold a cable. Another capability is to install a single cable or a bundle of small ducts into an occupied duct. The most expensive activity in installing a network is the need for civil works. Thus, re-using ducts occupied with one cable, leaving some space, is a tempting and often possible and cost-effective alternative.

HISTORY

The technique of installing flexible and lightweight fibre optic units using compressed air was developed during the 1980s by British Telecom. This early version of jetting did not use additional pushing. True cable jetting was invented by Willem Griffioen of KPN Research in the late 1980s. The necessary equipment was developed in cooperation with Plumettaz, Switzerland.

OPTICAL FIBRE CABLE

An optical fibre cable is a cable containing one or more optical fibers that are used to carry light. The optical fibre elements are typically individually coated with plastic layers and contained in a protective tube suitable for the environment where the cable will be deployed. Different types of cable are used for different applications, for example long distance telecommunication, or providing a high-speed data connection between different parts of a building.

DESIGN

Optical fibre consists of a core and a cladding layer, selected for total internal reflection due to the difference in the refractive index between the two. In practical fibers, the cladding is usually coated with a layer of acrylate polymer or polyimide. This coating protects the fibre from damage but does not contribute to its optical waveguide properties. Individual coated fibers (or fibers formed into ribbons or bundles) then have a tough resin buffer layer and/or core tube(s) extruded around them to form the cable core. Several layers of protective sheathing, depending on the application, are added to form the cable. Rigid fibre assemblies sometimes put light-absorbing ("dark") glass between the fibers, to prevent light that leaks out of one fibre from entering another. This reduces cross-talk between the fibers, or reduces flare in fibre bundle imaging applications.For indoor applications, the jacketed fibre is generally enclosed, with a bundle of flexible fibrous polymer *strength members* like aramid (*e.g.*. Twaron or Kevlar), in a lightweight plastic cover to form a simple cable. Each end of the cable may be terminated with a specialized optical fibre connector to allow it to be easily connected and disconnected from transmitting and receiving equipment.

For use in more strenuous environments, a much more robust cable construction is required. In *loose-tube construction* the fibre is laid helically into semi-rigid tubes, allowing the cable to stretch without stretching the fibre itself. This protects the fibre from tension during laying and due to temperature changes. Loose-tube fibre may be "dry block" or gel-filled. Dry block offers less protection to the fibers than gel-filled, but costs considerably less. Instead of a loose tube, the fibre may be embedded in a heavy polymer jacket, commonly called "tight buffer" construction. Tight buffer cables are offered for a variety of applications, but the two most common are "Breakout" and "Distribution". Breakout cables normally contain a ripcord, two non-conductive dielectric strengthening members (normally a glass rod epoxy), an aramid yarn, and 3 mm buffer tubing with an additional layer of Kevlar surrounding each fibre. The ripcord is a parallel cord of strong yarn that is situated under the jacket(s) of the cable for jacket removal. Distribution cables have an overall Kevlar wrapping, a ripcord, and a 900 micrometer buffer coating surrounding each fibre. These *fibre units* are commonly bundled with additional steel strength members, again with a helical twist to allow for stretching.

A critical concern in outdoor cabling is to protect the fibre from contamination by water. This is accomplished by use of solid barriers such as copper tubes, and water-repellent jelly or water-absorbing powder surrounding the fibre.

Finally, the cable may be armored to protect it from environmental hazards, such as construction work or gnawing animals. Undersea cables are more heavily armored in their near-shore portions to protect them from boat anchors, fishing gear, and even sharks, which may be attracted to the electrical power that is carried to power amplifiers or repeaters in the cable.

Modern cables come in a wide variety of sheathings and armor, designed for applications such as direct burial in trenches, dual use as power lines, installation in conduit, lashing to aerial telephone poles, submarine installation, and insertion in paved streets.

CAPACITY AND MARKET

In September 2012, NTT Japan demonstrated a single fibre cable that was able to transfer 1 petabit per second (10bits/s) over a distance of 50 kilometers. Modern fibre cables can contain up to a thousand fibers in a single cable, with potential bandwidth in the terabytes per second. In some cases, only a small fraction of the fibers in a cable may be actually "lit". Companies can lease or sell the unused fibre to other providers who are looking for service in or through an area. Companies may "overbuild" their networks for the specific purpose of having a large network of dark fibre for sale, reducing the overall need for trenching and municipal permitting.

RELIABILITY AND QUALITY

Optical fibers are very strong, but the strength is drastically reduced by unavoidable microscopic surface flaws inherent in the manufacturing process. The initial fibre strength, as well as its change with time, must be considered relative to the stress imposed on the fibre during handling, cabling, and installation for a given set of environmental conditions. There are three basic scenarios that can lead to strength degradation and failure by inducing flaw growth: dynamic fatigue, static fatigues, and zero-stress aging. Telcordia GR-20, *Generic Requirements for Optical Fibre and Optical Fibre Cable*, contains reliability and quality criteria to protect optical fibre in all operating conditions. The criteria concentrate on conditions in an outside plant (OSP) environment.

For the indoor plant, similar criteria are in Telcordia GR-409, *Generic Requirements for Indoor Fibre Optic Cable*.

JACKET MATERIAL

The jacket material is application specific. The material determines the mechanical robustness, aging due to UV radiation, oil resistance, etc. Nowadays PVC is being replaced by halogen free alternatives, mainly driven by more stringent regulations.

Material	Halogen-free	UV Resistance	Remark
LSFH Polymer	Yes	Good	Good for indoor use
Polyvinyl chloride (PVC)	N–	Good	Being replaced by LSFH Polymer
Polyethylene (PE)	Yes	Poor	Good for outdoor applications
Polyurethane (PUR)	Yes	?	Highly flexible cables
Polybutylene terephthalate (PBT)	Yes	Fair?	Good for indoor use
Polyamide (PA)	Yes	Good-Poor	Indoor and outdoor use

FIBRE MATERIAL

There are two main types of material used for optical fibers. These are glass and plastic. They offer widely different characteristics and therefore fibers made from the two different substances find uses in very different applications.

COLOUR CODING

Patch cords

The buffer or jacket on patchcords is often colour-coded to indicate the type of fibre used. The strain relief "boot" that protects the fibre from bending at a connector is colour-coded to indicate the type of connection. Connectors with a plastic shell (such as SC connectors) typically use a colour-coded shell. Standard colour codings for jackets and boots (or connector shells) are shown below:

Buffer/jacket colour	Meaning
Orange	multi-mode optical fibre
Aqua	OM3 or OM4 10 gig laser-optimized 50/125 micrometer multi-mode optical fibre
Violet	OM4 multi-mode optical fibre (some vendors)
Grey	outdated colour code for multi-mode optical fibre
Yellow	single-mode optical fibre
Blue	Sometimes used to designate polarization-maintaining optical fibre

Connector Boot	Meaning	Comment
Blue	Physical Contact (PC), 0°	mostly used for single mode fibers; some manufacturers use this for polarization-maintaining optical fibre.
Green	Angle Polished (APC), 8°	

Black	Physical Contact (PC), 0°	
Grey, Beige	Physical Contact (PC), 0°	multimode fibre connectors
White	Physical Contact (PC), 0°	
Red		High optical power. Sometimes used to connect external pump lasers or Raman pumps.

Remark: It is also possible that a small part of a connector is additionally colour-coded, *e.g.*. the leaver of an E-2000 connector or a frame of an adapter. This additional colour coding indicates the correct port for a patchcord, if many patchcords are installed at one point.

Multi-fibre cables

Individual fibers in a multi-fibre cable are often distinguished from one another by colour-coded jackets or buffers on each fibre. The identification scheme used by Corning Cable Systems is based on EIA/TIA-598, "Optical Fibre Cable Colour Coding." EIA/TIA-598 defines identification schemes for fibers, buffered fibers, fibre units, and groups of fibre units within outside plant and premises optical fibre cables. This standard allows for fibre units to be identified by means of a printed legend. This method can be used for identification of fibre ribbons and fibre subunits. The legend will contain a corresponding printed numerical position number and/or colour for use in identification.

EIA598-A Fibre Colour Chart — Colour coding of Premises Fibre Cable

Position	Jacket color	Position	Jacket colour
1	blue	13	blue/black
2	orange	14	orange/black
3	green	15	green/black
4	brown	16	brown/black
5	slate	17	slate/black
6	white	18	white/black
7	red	19	red/black
8	black	20	black/yellow
9	yellow	21	yellow/black
10	violet	22	violet/black
11	rose	23	rose/black
12	aqua	24	aqua/black

Fibre Type / Class	Diameter (μm)	Jacket Colour
Multimode 1a	50/125	range
Multimode 1a	62.5/125	Slate
Multimode 1a	85/125	Blue
Multimode 1a	100/140	Green
Singlemode IVa	All	Yellow
Singlemode IVb	All	Red

PROPAGATION SPEED AND DELAY

Optical cables transfer data at the speed of light in glass (slower than vacuum). This is typically around 180,000 to 200,000 km/s, resulting in 5.0 to

5.5 microseconds of latency per km. Thus the round-trip delay time for 1000 km is around 11 milliseconds.

LOSSES

Typical modern multimode graded-index fibers have 3 dB/km of attenuation loss at 850 nm and 1 dB/km at 1300 nm. 9/125 singlemode loses 0.4/0.25 dB/km at 1310/1550 nm. POF (plastic optical fibre) loses much more: 1 dB/m at 650 nm. Plastic optical fibre is large core (about 1mm) fibre suitable only for short, low speed networks such as within cars.

Each connection made adds about 0.6 dB of average loss, and each joint (splice) adds about 0.1 dB. Depending on the transmitter power and the sensitivity of the receiver, if the total loss is too large the link will not function reliably.

Invisible IR light is used in commercial glass fibre communications because it has lower attenuation in such materials than visible light. However, the glass fibers will transmit visible light somewhat, which is convenient for simple testing of the fibers without requiring expensive equipment. Splices can be inspected visually, and adjusted for minimal light leakage at the joint, which maximizes light transmission between the ends of the fibers being joined.

The charts at "Understanding wavelengths In fibre optics" and "Optical power loss (attenuation) in fibre" illustrate the relationship of visible light to the IR frequencies used, and show the absorption water bands between 850, 1300 and 1550 nm.

SAFETY

Because the infrared light used in communications can not be seen, there is a potential laser safety hazard to technicians. In some cases the power levels are high enough to damage eyes, particularly when lenses or microscopes are used to inspect fibers which are inadvertently emitting invisible IR. Inspection microscopes with optical safety filters are available to guard against this. Small glass fragments can also be a problem if they get under someone's skin, so care is needed to ensure that fragments produced when cleaving fibre are properly collected and disposed of appropriately.

HYBRID CABLES

There are hybrid optical and electrical cables that are used in wireless outdoor Fibre To The Antenna (FTTA) applications. In these cables, the optical fibers carry information, and the electrical conductors are used to transmit power. These cables can be placed in several environments to serve antennas mounted on poles, towers, and other structures.

According to Telcordia GR-3173, *Generic Requirements for Hybrid Optical and Electrical Cables for Use in Wireless Outdoor Fibre To The Antenna (FTTA)*

Applications, these hybrid cables have optical fibers, twisted pair/quad elements, coaxial cables and/or current-carrying electrical conductors under a common outer jacket. The power conductors used in these hybrid cables are for directly powering an antenna or for powering tower-mounted electronics exclusively serving an antenna. They have a nominal voltage normally less than 60 VDC or 108/120 VAC. Other voltages may be present depending on the application and the relevant National Electrical Code (NEC).

These types of hybrid cables may also be useful in other environments such as Distributed Antenna System (DAS) plants where they will serve antennas in indoor, outdoor, and roof-top locations. Considerations such as fire resistance, Nationally Recognized Testing Laboratory (NRTL) Listings, placement in vertical shafts, and other performance-related issues need to be fully addressed for these environments.

Since the voltage levels and power levels used within these hybrid cables vary, electrical safety codes consider the hybrid cable to be a power cable, which needs to comply with rules on clearance, separation, etc.

INNERDUCTS

Innerducts are installed in existing underground conduit systems to provide clean, continuous, low-friction paths for placing optical cables that have relatively low pulling tension limits. They provide a means for subdividing conventional conduit that was originally designed for single, large-diameter metallic conductor cables into multiple channels for smaller optical cables.

Types

Innerducts are typically small-diameter, semi-flexible subducts. According to Telcordia GR-356, there are three basic types of innerduct: smoothwall, corrugated, and ribbed. These various designs are based on the profile of the inside and outside diameters of the innerduct. The need for a specific characteristic or combination of characteristics, such as pulling strength, flexibility, or the lowest coefficient of friction, dictates the type of innerduct required.

Beyond the basic profiles or contours (smoothwall, corrugated, or ribbed), innerduct is also available in an increasing variety of multiduct designs. Multiduct may be either a composite unit consisting of up to four or six individual innerducts that are held together by some mechanical means, or a single extruded product having multiple channels through which to pull several cables. In either case, the multiduct is coilable, and can be pulled into existing conduit in a manner similar to that of conventional innerduct.

Placement

Innerducts are primarily installed in underground conduit systems that provide connecting paths between manhole locations. In addition to placement in conduit, innerduct can be directly buried, or aerially installed by lashing the innerduct to a steel suspension strand.

As stated in GR-356, cable is typically placed into innerduct in one of three ways. It may be

1. Pre-installed by the innerduct manufacturer during the extrusion process,
2. Pulled into the innerduct using a mechanically assisted pull line, or
3. Blown into the innerduct using a high air volume cable blowing apparatus.

CHIRAL PHOTONICS

Chiral Photonics, Inc. is a photonics company based in Pine Brook, New Jersey, founded in 1999. The company is developing a new class of optical devices based on twisting glass optical fibers. These in-fibre devices aim to displace discrete optical elements such as lasers, filters and sensors. They benefit from optical fiber's transmission efficiency, robustness and ease of integration.

The company hopes that its manufacturing process, which is completely automated and scalable, will result, for example, in communications lasers that are fraction of the cost and three times more efficient than today's semiconductor lasers. Chiral Photonics is also developing chirality in polymeric thin films which, for instance, would enable high quality projection displays.

FUNDING

Chiral Photonics had received funding from venture capital, angel, and government sources including a US$2 million National Institute of Standards and Technology Advanced Technology Program award in 2004.

TECHNOLOGY

Chiral Photonics' technology is an outgrowth of the 1997 discovery by two of the company's co-founders, Azriel Genack and Victor Kopp, that lasing in cholesteric liquid crystal (CLC) films is a result of their unique self-assembling helical (chiral) microstructure. CLCs are the thin-film material often used to fabricate fish tank thermometers or mood rings, that change colour with temperature changes. They change color because their molecules are arranged in a helical or chiral arrangement and with temperature the pitch of that helical structure changes, reflecting different wavelengths of light.

Drs. Genack and Kopp decided to pursue the possibility that CLCs, with their natural chiral structure, could provide a platform for a versatile new class of photonic devices. In biomimetic fashion, Chiral Photonics has abstracted the self-assembled structure of the organic CLCs to produce analogous optical devices using tiny lengths of inorganic, twisted fibre. Designing novel microforming towers, the company is able to fabricate devices based on fibers that can be twisted through more than 25,000 revolutions over a one-inch length.

These revolutions function as would a fibre Bragg grating. The density of twists per inch, or periodicity, demonstrably results in the light being coupled to the fibre cladding, scattered out of the fibre, or reflected back within the fibre. This interaction with light can be harnessed to produce sensors, polarizers/isolators, and filter/lasers, respectively.

APPLICATIONS

These basic components can be used for a variety of applications and all share a common production platform. In his comments upon the grant's award, William Sargeant, the National Science Foundation program officer who oversaw Chiral Photonics' first SBIR award, noted the range of existing and incipient markets. "This technology could be one of the most significant recent advances in the field of polarization and wavelength control. There is an enormous host of applications for which chiral fibre gratings could find markets."

Chiral Photonics' components are all of the all-fibre variety. Released products include linear and circular polarizers, ultra-high temperature sensors, customized harsh environment pressure, axial rotation, and liquid level sensors, and a spot size converting interconnect.

The spot size converter, while not of chiral geometry, leverages the company's glass microfabrication knowhow. The spot size converter (SSC) couples light between widely disparate (NA and MFD) components, such as, between <25 micrometre planar waveguides or laser diodes, and standard 125 micrometre SMF fibre. The SSC allows for direct light coupling with no air gap, sub-0.5dB loss, and extinction ratios of >20 dB for use in silicon photonics and other applications.

Chiral Photonics also offers twisted capillary tubes for proteomic analysis. The protein unfolds as it passes through the channel allowing for imaging. In other uses the rotation of the protein as it passes through the channel facilitates 360° imaging. The capillary tubes also have other microfluidic applications including mixing and uniform heat exchange.

PATENTS

Chiral Photonics has been issued 19 United States or International patents relating to its photonics research.

CHROMO-MODAL DISPERSION

Chromo-modal dispersion (CMD) results from exciting various modes of a multimode waveguide with unique spectral components of a broadband optical signal. Modal dispersion during propagation in the waveguide then provides group velocity dispersion to the signal. The large modal dispersion inherent to multimode waveguides enables the dispersion per unit length of a chromo-modal dispersion device to be several orders of magnitude higher than that of diffraction grating or dispersion compensating fibre-based dispersive elements.

APPLICATIONS

The ability to control chromatic dispersion is paramount in applications where the optical pulsewidth is critical, such as chirped pulse amplification and fibre optic communications.

OTHER DEVICES

Typically, devices used to generate large amounts (>100 ps/nm) of chromatic dispersion are based on diffraction gratings, chirped fibre Bragg gratings, or dispersion compensating fibre. Unfortunately, these dispersive elements suffer from one or more of the following restrictions:

1. Limited operational bandwidth
2. Limited total dispersion
3. Low peak power handling
4. Large spatial footprint.

CONSTRUCTION

The chromo-modal dispersion device is constructed by combining the angular dispersion of diffraction gratings with the modal dispersion of a multimode waveguide.

ADVANTAGES

The large dispersion and small footprint of the device make the chromo-modal dispersion device potentially useful for on-chip dispersion compensation using optical components such as integrated gratings and planar multimode waveguides. The advantage of physical compactness, combined with the magnitude and tunability of its dispersion suggest its potential use as a versatile tool for pulse stretching or compression in a variety of applications in which the capabilities of singlemode fibre or diffraction grating-based dispersive elements will not suffice.

CLADDING (FIBRE OPTICS)

Cladding is one or more layers of materials of lower refractive index, in intimate contact with a core material of higher refractive index. The cladding causes light to be confined to the core of the fibre by total internal reflection at the boundary between the two. Light propagation in the cladding is suppressed in typical fibre. Some fibers can support cladding modes in which light propagates in the cladding as well as the core. (From Federal Standard 1037C and from MIL-STD-188)

The numerical aperture of a fibre is a function of the indices of refraction of the cladding and the core by:

$$NA = \sqrt{n_{\text{core}}^2 - n_{\text{clad}}^2}$$

CLADDING MODE

In fibre optics, a cladding mode is a mode that is confined to the cladding of an optical fibre by virtue of the fact that the cladding has a higher refractive index than the surrounding medium, which is either air or the primary polymer overcoat. These modes are generally undesired. Modern fibers have a primary polymer overcoat with a refractive index that is slightly higher than that of the cladding, so that light propagating in the cladding is rapidly attenuated and disappears after only a few centimeters of propagation. An exception to this is double-clad fibre, which is designed to support a mode in its inner cladding, as well as one in its core.

CLEAVE (FIBRE)

A cleave in an optical fibre is a deliberate, controlled break, intended to create a perfectly flat endface, perpendicular to the longitudinal axis of the fibre. Since there are no crystalline planes in glass, this process is not cleavage in the crystallographic sense of the word, although the techniques used and the finished result are quite similar.

A good cleave is required for a successful splice of an optical fibre, whether by fusion or mechanical means. Also, some types of fibre-optic connectors do not employ abrasives and polishers. Instead, they use some type of cleaving technique to trim the fibre to its proper length, and produce a smooth, flat perpendicular endface.

PROCESS

A cleave is made by first introducing a microscopic fracture ("nick") into the fibre with a special tool, called a cleaving tool, which has a sharp blade of some hard material, such as diamond, sapphire, or tungsten carbide. If proper tension is applied to the fibre as the nick is made, or immediately afterwards (this may be done by the cleaving tool in some designs, or manually in other designs), the fracture will propagate in a controlled fashion, creating the desired endface.

TOOLS

- Pen-shaped scribe (aka diamond-tip scribe or diamond wedge scribe) looks like a ballpoint pen, but has a small wedge tip made of diamond or other hard material. This tool is used with the "scratch and pull" technique. First the fibre is scribed perpendicular to its length. The fibre is then pulled, which breaks at the scribe. This tool requires an experienced operator to produce good cleaves.

- Mechanical cleavers clamp the fibre in the correct position before a diamond wheel or blade scribes the fibre. Then, a force is applied and the fibre gives a nice break at the scribe. Mechanical cleavers give nicer and more repeatable cleaves.
- Multifiber cleavers are used for ribbon fibre cables.

CONCENTRICITY ERROR

The concentricity error of an optical fibre is the distance between the center of the two concentric circles that specify the cladding diameter and the center of the two concentric circles that specify the core diameter. The concentricity error is used in conjunction with tolerance fields to specify or characterize optical fibre core and cladding geometry.

CORE (OPTICAL FIBRE)

The core of a conventional optical fibre is a cylinder of glass or plastic that runs along the fiber's length. The core is surrounded by a medium with a lower index of refraction, typically a cladding of a different glass, or plastic. Light travelling in the core reflects from the core-cladding boundary due to total internal reflection, as long as the angle between the light and the boundary is less than the critical angle. As a result, the fibre transmits all rays that enter the fibre with a sufficiently small angle to the fiber's axis. The limiting angle is called the acceptance angle, and the rays that are confined by the core/cladding boundary are called guided rays. The core is characterized by its diameter or cross-sectional area. In most cases the core's cross-section should be circular, but the diameter is more rigorously defined as the average of the diameters of the smallest circle that can be circumscribed about the core-cladding boundary, and the largest circle that can be inscribed within the core-cladding boundary. This allows for deviations from circularity due to manufacturing variation. Another commonly quoted statistic for core size is the mode field diameter. This is the diameter at which the intensity of light in the fibre falls to some specified fraction of maximum (usually 1/eH"13.5 per cent). For single-mode fibre, the mode field diameter is larger than the physical diameter of the core, because the light penetrates slightly into the cladding as an evanescent wave.

COUPLING LOSS

Coupling loss also known as connection loss is the loss that occurs when energy is transferred from one circuit, circuit element, or medium to another. Coupling loss is usually expressed in the same units—such as watts or decibels—as in the originating circuit element or medium. Coupling loss in fibre optics refers to the power loss that occurs when coupling light from one optical device or medium to another. (See also Optical return loss.) Coupling losses can result from a number of factors. In electronics (*see Coupling (electronics)*), impedance mismatch between coupled components results in a reflection of a portion of the energy at the interface. Likewise, in optical systems, where there is a change in index of refraction (most commonly at a fibre/air interface), a portion of the energy is reflected back into the source component.

Another major source of optical coupling loss is geometrical. As an example, two fibers coupled end-to-end may not be precisely aligned, with the result that the two cores overlap somewhat. Light exiting the source fibre at a portion of its core that is not aligned with the core of the receiving fibre will not (in general) be coupled into the second fibre. While some such light will be coupled into the second fibre, it is not likely to be efficiently coupled, nor will it generally travel in an appropriate mode in the second fibre. Similarly, even for two perfectly aligned cores, where there is a gap of any significant distance between the two fibers, there will be some geometric loss due to spread of the beam. Some percentage of the light rays exiting the source fibre face will not intersect the second fibre within its entrance cone.

CUTBACK TECHNIQUE

In telecommunications, a cutback technique is a destructive technique for determining certain optical fibre transmission characteristics, such as attenuation and bandwidth.

PROCEDURE

The measurement technique consists of:

1. Performing the desired measurements on a long length of the fibre under test,
2. Cutting the fibre under test at a point near the launching end,
3. Repeating the measurements on the short length of fibre, and
4. Subtracting the results obtained on the short length to determine the results for the residual long length.

The cut should be made to retain 1 meter or more of the fibre, in order to establish equilibrium mode distribution conditions for the second measurement. In a multimode fibre, the lack of an equilibrium mode distribution could introduce errors in the measurement due to output coupling effects. In a single-mode fibre, measuring a shorter cutback fibre could result in significant transmission of cladding modes (light carried in the cladding rather than the core of the optical fibre), distorting the measurement. The errors introduced will result in conservative results (*i.e.*, higher transmission losses and lower bandwidths) than would be realized under equilibrium conditions.

BENEFITS

The benefit of this technique is that it allows measurement of the fibre characteristics without introducing errors due to variation in the launch conditions. For example, the coupling efficiency of the light source is kept consistent between the initial and the cutback measurements. Several characteristics may be determined using the same test fibre.

ATTENUATION MEASUREMENT

Since the attenuation is defined as proportional to the logarithm of the ratio between $P(x)$ and $P(y)$, where P is the power at point x and y respectively. Using the cutback technique, the power transmitted through a fibre of known length is measured and compared with the same measurement for the same fibre cut to a length of $2m$ approximately.

RELATED TECHNIQUES

A variation of the cutback technique is the substitution method, in which measurements are made on a full length of fibre, and then on a short length of fibre having the same characteristics (core size, numerical aperture), with the results from the short length being subtracted to give the results for the full length.

DISTRIBUTED BRAGG REFLECTOR

A distributed Bragg reflector (DBR) is a reflector used in waveguides, such as optical fibers. It is a structure formed from multiple layers of alternating materials with varying refractive index, or by periodic variation of some characteristic (such as height) of a dielectric waveguide, resulting in periodic variation in the effective refractive index in the guide. Each layer boundary causes a partial reflection of an optical wave. For waves whose wavelength is close to four times the optical thickness of the layers, the many reflections combine with constructive interference, and the layers act as a high-quality reflector. The range of wavelengths that are reflected is called the photonic stopband. Within this range of wavelengths, light is "forbidden" to propagate in the structure.

REFLECTIVITY

The DBR's reflectivity, R, for intensity is approximately given by

$$R = \left[\frac{n_o (n_2)^{2N} - n_s (n_1)^{2N}}{n_o (n_2)^{2N} + n_s (n_1)^{2N}} \right]^2$$

where n_0, n_1, n_2 and n_s are the respective refractive indices of the originating medium, the two alternating materials, and the terminating medium (*i.e.*. backing or substrate); and N is the number of repeated pairs of low/high refractive index material.

The bandwidth $\Delta\lambda_0$ of the photonic stopband can be calculated by

$$\Delta\lambda_0 = \frac{4\lambda_0}{\pi}\arcsin\left(\frac{n_2 - n_1}{n_2 + n_1}\right),$$

where λ_0 is the central wavelength of the band.

Increasing the number of pairs in a DBR increases the mirror reflectivity and increasing the refractive index contrast between the materials in the Bragg pairs increases both the reflectivity and the bandwidth. A common choice of materials for the stack is titanium dioxide ($n \approx 2.5$) and silica ($n \approx 1.5$). Substituting into the formula above gives a bandwidth of about 200 nm for 630 nm light.

Distributed Bragg reflectors are critical components in vertical cavity surface emitting lasers and other types of narrow-linewidth laser diodes such as distributed feedback (DFB) lasers and distributed bragg reflector (DBR) lasers. They are also used to form the cavity resonator (or optical cavity) in fibre lasers and free electron lasers.

TE and TM mode reflectivity

This part discusses the interaction of transverse electric (TE) and transverse magnetic (TM) polarized light with the DBR structure, over several wavelengths and incidence angles. This reflectivity of the DBR structure (described below) was calculated using the transfer-matrix method (TMM), where the TE mode alone is highly reflected by this stack, while the TM modes are passed through. This also shows the DBR acting as a polarizer. For TE and TM incidence we have the reflection spectra of a DBR stack, corresponding to a 6 layer stack of dielectric contrast of 11.5, between an air and dielectric layers. The thicknesses of the air and dielectric layers are 0.8 and 0.2 of the period, respectively. The wavelength in the figures below, corresponds to multiples of the cell period. This DBR is also a simple example of a 1D photonic crystal. It has a complete TE band gap, but only a pseudo TM band gap.

DISTRIBUTED TEMPERATURE SENSING

Distributed temperature sensing systems (DTS) are optoelectronic devices which measure temperatures by means of optical fibres functioning as linear sensors. Temperatures are recorded along the optical sensor cable, thus not at points, but as a continuous profile. A high accuracy of temperature determination is achieved over great distances. Typically the DTS systems can locate the temperature to a spatial resolution of 1 m with accuracy to within ±1°C at a resolution of 0.01°C. Measurement distances of greater than 30 km can be monitored and some specialised systems can provide even tighter spatial resolutions.

MEASURING PRINCIPLE—RAMAN EFFECT

Physical measurement dimensions, such as temperature or pressure and tensile forces, can affect glass fibres and locally change the characteristics of light transmission in the fibre. As a result of the damping of the light in the quartz glass fibres through scattering, the location of an external physical effect can be determined so that the optical fibre can be employed as a linear sensor. Optical fibres are made from doped quartz glass. Quartz glass is a form of silicon dioxide (SiO_2) with amorphous solid structure. Thermal effects induce lattice oscillations within the solid. When light falls onto these thermally excited molecular oscillations, an interaction occurs between the light particles (photons) and the electrons of the molecule. Light scattering, also known as Raman scattering, occurs in the optical fibre. Unlike incident light, this scattered light undergoes a spectral shift by an amount equivalent to the resonance frequency of the lattice oscillation. The light scattered back from the fibre optic therefore contains three different spectral sharesf :

- The Rayleigh scattering with the wavelength of the laser source used,
- The Stokes line components from photons shifted to longer wavelength (lower frequency), and
- The anti-Stokes line components with photons shifted to shorter wavelength (higher frequency) than the Rayleigh scattering.

The intensity of the so-called anti-Stokes band is temperature-dependent, while the so-called Stokes band is practically independent of temperature. The local temperature of the optical fibre is derived from the ratio of the anti-Stokes and Stokes light intensities.

MEASURING PRINCIPLE—OTDR AND OFDR TECHNOLOGY

There are two basic principles of measurement for distributed sensing technology, OTDR (Optical Time Domain Reflectometry) and OFDR (Optical Frequency Domain Reflectometry). For Distributed Temperature Sensing often a Code Correlation technology is employed which carries elements from both principles.

OTDR was developed more than 20 years ago and has become the industry standard for telecom loss measurements which detects the—compared to Raman signal very dominant—Rayleigh backscattering signals. The principle for OTDR is quite simple and is very similar to the time of flight measurement used for radar. Essentially a narrow laser pulse generated either by semiconductor or solid state lasers is sent into the fibre and the backscattered light is analysed. From the time it takes the backscattered light to return to the detection unit it is possible to locate the location of the temperature event.

Alternative DTS evaluation units deploy the method of Optical Frequency Domain Reflectometry (OFDR). The OFDR system provides information on the local characteristic only when the backscatter signal detected during the entire measurement time is measured as a function of frequency in a complex fashion, and then subjected to Fourier transformation. The essential principles of OFDR technology are the quasi continuous wave mode employed by the laser and the narrow-band detection of the optical back scatter signal. This is offset by the technically difficult measurement of the Raman scatter light and rather complex signal processing, due to the FFT calculation with higher linearity requirements for the electronic components.

Code Correlation DTS sends on/off sequences of limited length into the fibre. The codes are chosen to have suitable properties, *e.g.*. Binary Golay code. In contrast to OTDR technology, the optical energy is spread over a code rather than packed into a single pulse. Thus a light source with lower peak power compared to OTDR technology can be used, *e.g.*. long life compact semiconductor lasers. The detected backscatter needs to be transformed—similar to OFDR technology—back into a spatial profile, *e.g.*. by cross-correlation. In contrast to OFDR technology, the emission is finite (for example 128 bit) which avoids that weak scattered signals from far are superposed by strong scattered signals from short distance, improving the Shot noise and the signal-to-noise ratio.Using these techniques it is possible to analyse distances of greater than 30 km from one system and to measure temperature resolutions of less than 0.01°C.

CONSTRUCTION OF SENSING CABLE AND SYSTEM INTEGRATION

The temperature measuring system consists of a controller (laser source, pulse generator for OTDR or code generator for Code Correlation or modulator and HF mixer for OFDR, optical module, receiver and micro-processor unit) and a quartz glass fibre as line-shaped temperature sensor. The fibre optic cable (can be 30 km+ in length) is passive in nature and has no individual sensing points and therefore can be manufactured based on standard telecoms fibres. This offers excellent economies of scale. Because the system designer/integrator does not have to worry about the precise location of each sensing point the cost for designing and installing a sensing system based on distributed fibre optic sensors is greatly reduced from that of traditional sensors. Additionally, because the sensing cable has no moving parts and design lives of 30 years +, the maintenance and operation costs are also considerably less than for conventional sensors. Additional benefits of fibre optic sensing technology are that it is immune to electromagnetic interference, vibration and is safe for use in hazardous zones (the laser power falls below the levels that can cause ignition), thus making these sensors ideal for use in industrial sensing applications.

With regards to the construction of the sensing cable, although it is based on standard fibre optics, care must be taken in the design of the individual sensing cable to ensure that adequate protection is provided for the fibre. This must take into account operating temperature (standard cables operate to 85°C but it is possible to measure up to 700°C with the correct design), gaseous environment (hydrogen can cause deterioration of the measurement though "hydrogen darkening" - aka attenuation - of the silica glass compounds) and mechanical protection.

Most of the available DTS systems have flexible system architectures and are relatively simple to integrate into industrial control systems such as SCADA. In the oil and gas industry an XML based file standard (WITSML) has been developed for transfer of data from DTS instruments. The standard is maintained by Energistics.

LASER SAFETY AND OPERATION OF SYSTEM

When operating a system based on optical measurements such as optical DTS, laser safety requirements need to be considered for permanent installations. Many systems use low power laser design, *e.g.*. with classification as laser safety class 1M, which can be applied by anyone (no approved laser safety officers required). Some systems are based on higher power lasers of a 3B rating, which although safe for use by approved laser safety officers, may not be suitable for permanent installations.

The advantage of purely passive optical sensor technology is the lack of electric or electromagnetic interaction. Some DTS systems on the market use a special low power design and are inherently safe in explosive environments, *e.g.*. certified to ATEX directive Zone 0.

For use in fire detection application, regulations usually require certified systems according to relevant standards, such as EN 54-5 or EN 54-22 (Europe), UL521 or FM (USA), cUL521 (Canada) and/or other national or local standards.

TEMPERATURE ESTIMATIONS BY USING DTS

Temperature distributions can be used to develop models based on the Proper Orthogonal Decomposition Method or principal component analysis. This allows to reconstruct the temperature distribution by measuring only in a few spatial locations

APPLICATIONS

Distributed temperature sensing can be deployed successfully in multiple industrial segments:

- Oil and gas production—permanent downhole monitoring, coil tubing optical enabled deployed intervention systems, slickline optical cable deployed intervention systems.
- Power cable and transmission line monitoring (ampacity optimisation)
- Fire detection in tunnels, industrial conveyor belts and special hazard buildings

- Industrial induction furnace surveillance
- Integrity of liquid natural gas (LNG) carriers and terminals
- Leakage detection at dikes and dams
- Temperature monitoring in plant and process engineering, including transmission pipelines
- Storage tanks and vessels

More recently, DTS has been applied for ecological monitoring as well:

- Stream temperature
- Groundwater source detection
- Temperature profiles in a mine shaft and over lakes and glaciers
- Deep rainforest ambient temperature at various foliage densities
- Temperature profiles in an underground mine, Australia
- Temperature profiles in ground loop heat exchangers (used for ground coupled heating and cooling systems)

DUKENET COMMUNICATIONS

Headquartered in Charlotte, N.C., DukeNet Communications primarily serves the southeastern United States with a robust metro and long haul fibre-optic network capable of delivering 100 Gbit/s services, enabling cloud computing and high-bandwidth applications for enterprise, data center, government entities and wholesale carrier customers.

Formed in 1994 as the telecommunications arm of Duke Energy, DukeNet primarily offered wholesale fibre transport services until moving into the newly developing Fibre to the Tower (FTTT) space in 2006. Today DukeNet is among the top 10 FTTT providers in the United States. In December 2010, Duke Energy and Alinda Capital Partners LLC formed a joint venture to grow DukeNet Communications in the enterprise and data center space. On October 7, 2013, Time Warner Cable announced that it had agreed to acquire DukeNet Communications LLC for $600 million, and on January 6, 2014, the deal officially closed.

DUKENET'S SERVICES

DukeNet provides bandwidth Infrastructure and carrier-neutral colocation and interconnection services, provided over regional and metropolitan fibre networks, enabling the transport of data and network interconnection.

SERVICES SUMMARY

- Carrier Ethernet
- Private Line
- IP Services/DIA
- Managed Services
- DAS/Small Cell

- Cell Site Backhaul/FTTT
- Dark Fibre
- Colocation
- Custom Network Solutions
- Professional Services

INDUSTRIES SERVED

- Enterprise/ Commercial
- Financialf
- Government/Education(5)
- Health care
- Manufacturing
- Technology
- Data Center
- Wireless
- Carrier

EFFECTIVE MODE VOLUME

For an optical fibre, the effective mode volume is the square of the product of the diameter of the near-field pattern and the sine of the radiation angle of the far-field pattern. The diameter of the near-field radiation pattern is defined here as the full width at half maximum and the radiation angle at half maximum radiant intensity. Effective mode volume is proportional to the breadth of the relative distribution of power amongst the modes in a multimode fibre. It is not truly a spatial volume but rather an *"optical volume"* equal to the product of area and solid angle. The power divided by the effective mode volume is proportional to the radiance of the light emitted by the fibre.

EQUILIBRIUM MODE DISTRIBUTION

The equilibrium mode [power] distribution of light travelling in an optical waveguide or fibre, is the distribution of light that is no longer changing with fibre length or with input modal excitation. This phenomenon requires both mode filtering and mode mixing to occur in the fibre to produce a state that is independent of the mode power distribution launched by the light source. At propagation distances exceeding the equilibrium length, intramodal pulse distortion increases (bandwidth decreases) as the square root of length.

The term *equilibrium length* is sometimes used to describe a *stationary* mode distribution, which is the length of multi-mode optical fibre necessary to attain a static mode distribution from a specific excitation condition. Equilibrium length is, strictly, the longest such length, as would result from a widely variable range of input excitation. Other terms for equilibrium length are *equilibrium coupling length* and *equilibrium mode distribution length*.

Equilibrium mode [power] distributions were reported in early multimode transmission systems at propagation distances as short as a few hundred metres. However, as fibre manufacturing improved, the minute waveguide dimensional and structural changes that produce mode-mixing have been greatly reduced. The length of fibre required to attain true equilibrium is now much greater than the length of practical multimode transmission systems, which makes the term effectively obsolete.

In the absence of strong mode-mixing, high order mode filtering is the primary remaining mechanism for potential change of an input mode power distribution. If a well-aligned laser is the optical source, the mode power distribution is highly concentrated in the lowest order modes, and remains essentially unchanged with distance due to the lack of mode-mixing. If an optical source that overfills the fibre is used, only the highest order guided mode group experiences excess attenuation, and the mode power distribution becomes slightly filtered as a result. (Mandrel wrapping is a viable method to artificially create this state.) Such mode power distributions are stationary; neither changes with fibre length, but equilibrium does not exist in either case because the distributions remain dependent on the input power distribution.

FANOUT CABLE

Breakout-style fiberoptic cable (also called breakout cable or fanout cable), is an optical fibre cable containing several jacketed simplex optical fibers packaged together inside an outer jacket. This differs from distribution-style cable, in which tight-buffered fibers are bundled together, with only the outer cable jacket of the cable protecting them. The design of breakout-style cable adds strength for ruggedized drops, however the cable is larger and more expensive than distribution-style cable. Breakout cable is suitable for short riser and plenum applications and also for use in conduits, where a very simple cable run is planned to avoid the use of any splicebox or spliced fibre pigtails.

Because each fibre is individually reinforced, the cable can be easily divided into individual fibre lines. Each simplex cable within the outer jacket may be broken out and then continue as a patch cable, for example in a fibre to the desk application in an office building. This enables connector termination without requiring special junctions, and can reduce or eliminate the need for fiberoptic patch panels or an optical distribution frame. Breakout cable requires terminations to be done with simple connectors, which may be preferred for some situations. A more common solution today is the use of a fanout kit that adds a jacket to the very fine strands of other cable types.

FIBRE LASER

A fibre laser or fibre laser is a laser in which the active gain medium is an optical fibre doped with rare-earth elements such as erbium, ytterbium, neodymium, dysprosium, praseodymium, and thulium. They are related to doped fibre amplifiers, which provide light amplification without lasing. Fibre non-linearities, such as stimulated Raman scattering or four-wave mixing can also provide gain and thus serve as gain media for a fibre laser.

ADVANTAGES AND APPLICATIONS

The advantages of fibre lasers over other types include:

- Light is already coupled into a flexible fibre: The fact that the light is already in a fibre allows it to be easily delivered to a movable focusing element. This is important for laser cutting, welding, and folding of metals and polymers.
- High output power: Fibre lasers can have active regions several kilometers long, and so can provide very high optical gain. They can support kilowatt levels of continuous output power because of the fiber's high surface area to volume ratio, which allows efficient cooling.
- High optical quality: The fiber's waveguiding properties reduce or eliminate thermal distortion of the optical path, typically producing a diffraction-limited, high-quality optical beam.
- Compact size: Fibre lasers are compact compared to rod or gas lasers of comparable power, because the fibre can be bent and coiled to save space.
- Reliability: Fibre lasers exhibit high vibrational stability, extended lifetime, and maintenance-free turnkey operation.
- High peak power and nanosecond pulses enable effective marking and engraving.
- The additional power and better beam quality provide cleaner cut edges and faster cutting speeds.
- Lower cost of ownership.
- Fibre lasers are now being used to make high-performance surface-acoustic wave (SAW) devices. These lasers raise throughput and lower cost of ownership in comparison to older solid-state laser technology.

Fibre laser can also refer to the machine tool that includes the fibre resonator. Applications of fibre lasers include material processing (marking, engraving, cutting), telecommunications, spectroscopy, medicine, and directed energy weapons.

DESIGN AND MANUFACTURE

Unlike most other types of lasers, the laser cavity in fibre lasers is constructed monolithically by fusion splicing different types of fibre; fibre Bragg gratings replace conventional dielectric mirrors to provide optical feedback. Another type is the single longitudinal mode operation of ultra narrow distributed feedback lasers (DFB) where a phase-shifted Bragg grating overlaps the gain medium. Fibre lasers are pumped by semiconductor laser diodes or by other fibre lasers. Q-switched pulsed fibre lasers offer a compact, electrically efficient alternative to Nd:YAG technology.

Double-clad fibers

Many high-power fibre lasers are based on double-clad fibre. The gain medium forms the core of the fibre, which is surrounded by two layers of cladding. The lasing mode propagates in the core, while a multimode pump beam propagates in the inner cladding layer. The outer cladding keeps this pump light confined. This arrangement allows the core to be pumped with a much higher-power beam than could otherwise be made to propagate in it, and allows the conversion of pump light with relatively low brightness into a much higher-brightness signal. As a result, fibre lasers and amplifiers are occasionally referred to as "brightness converters." There is an important question about the shape of the double-clad fibre; a fibre with circular symmetry seems to be the worst possible design. The design should allow the core to be small enough to support only a few (or even one) modes. It should provide sufficient cladding to confine the core and optical pump section over a relatively short piece of the fibre.

Power scaling

Recent developments in fibre laser technology have led to a rapid and large rise in achieved diffraction-limited beam powers from diode-pumped solid-state lasers. Due to the introduction of large mode area (LMA) fibers as well as continuing advances in high power and high brightness diodes, continuous-wave single-transverse-mode powers from Yb-doped fibre lasers have increased from 100 W in 2001 to >20 kW. Commercial single-mode lasers have reached 10 kW in CW power. In 2014 a combined beam fibre laser demonstrated power of 30 kW.

Mode locking

Passive mode locking

Non-linear polarization rotation

When linearly polarized light is incident to a piece of weakly birefringent fibre, the polarization of the light will generally become elliptically polarized in the fibre. The orientation and ellipticity of the final light polarization is fully determined by the fibre length and its birefringence. However, if the intensity of the light is strong, the non-linear optical Kerr effect in the fibre must be considered, which introduces extra changes to the light polarization. As the polarization change introduced by the optical Kerr effect depends on the light intensity, if a polarizer is put behind the fibre, the light intensity transmission through the polarizer will become light intensity dependent. Through appropriately selecting the orientation of the polarizer or the length of the fibre, an artificial saturable absorber effect with ultra-fast response could then be achieved in such a system, where light of higher intensity experiences less absorption loss on the polarizer. The NPR technique makes use of this artificial saturable absorption to achieve the passive mode locking in a fibre laser. Once a mode-locked pulse is formed, the non-linearity of the fibre further shapes the pulse into an optical soliton and consequently the ultrashort soliton operation is obtained in the laser. Soliton operation is almost a generic feature of the fibre lasers mode-locked by this technique and has been intensively investigated.

Semiconductor saturable absorber mirrors (SESAMs)

Semiconductor saturable absorbers were used for laser mode-locking as early as 1974 when p-type germanium is used to mode lock a CO2 laser which generated pulses ~500 ps . Modern SESAMs are III-V semiconductor single quantum well (SQW) or multiple quantum wells grown on semiconductor distributed Bragg reflectors (DBRs). They were initially used in a Resonant Pulse Modelocking (RPM) scheme as starting mechanisms for Ti:Sapphire lasers which employed KLM as a fast saturable absorber . RPM is another coupled-cavity mode-locking technique. Different from APM lasers which employ non-resonant Kerr-type phase non-linearity for pulse shortening, RPM employs the amplitude non-linearity provided by the resonant band filling effects of semiconductors . SESAMs were soon developedinto intracavity saturable absorber devices because of more inherent simplicity with this structure. Since then, the use of SESAMs has enabled the pulse durations, average powers, pulse energies and repetition rates of ultrafast solid-state lasers to be improved by several orders of magnitude. Average power of 60 W and repetition rate up to 160 GHz were obtained. By using SESAM-assisted KLM, sub-6 fs pulses directly from a Ti: Sapphire oscillator was achieved. A major advantage SESAMs have over other saturable absorber techniques is that absorber parameters can be easily controlled over a wide range of values. For example, saturation fluence can be controlled by varying the reflectivity of the top reflector while modulation depth and recovery time can be tailored by changing the low temperature growing conditions for the absorber layers . This freedom of design has further extended the application of SESAMs into modelocking of fibre lasers where a relatively high modulation depth is needed to ensure self-starting and operation stability. Fibre lasers working at ~ 1 μm and 1.5 μm were successfully demonstrated.

Carbon nanotube saturable absorbers

- Graphene saturable absorbers: Graphene is a one-atom-thick planar sheet of sp2-bonded carbon atoms that are densely packed in a honeycomb crystal lattice. Optical absorption from graphene can become saturated when the input optical intensity is above a threshold value. This non-linear optical behaviour is termed saturable absorption and the threshold value is called the saturation fluency. Graphene can be saturated readily under strong excitation over the visible to near-infrared region, due to the universal optical absorption and zero band gap. This has relevance for the mode locking of fibre lasers, where wideband tunability may be obtained using graphene as the saturable absorber. Due to this special property, graphene has wide application in ultrafast photonics. Furthermore, comparing with the SWCNTs, as graphene has a 2D structure it should have much smaller non-saturable loss and much higher damage threshold. Self-started mode locking and stable soliton pulse emission with high energy have been achieved with a graphene saturable absorber in an erbium-doped fibre laser. Atomic layer graphene possesses wavelength-insensitive ultrafast saturable absorption, which can be exploited as a "full-band" mode locker. With an erbium-doped dissipative soliton fibre laser mode locked with few layer graphene, it has been experimentally shown that dissipative solitons with continuous wavelength tuning as large as 30 nm (1570–1600 nm) can be obtained.

Active mode locking

Active mode-locking is normally achieved by modulating the loss (or gain) of the laser cavity at a repetition rate equivalent to the cavity frequency, or a harmonic thereof. In practice, the modulator can be acousto-optic or electro-optic modulator, Mach-Zehnder integrated-optic modulators, or a semiconductor electro-absorption modulator (EAM). The principle of active mode-locking with a sinusoidal modulation. In this situation, optical pulses will form in such a way as to minimize the loss from the modulator. The peak of the pulse would automatically adjust in phase to be at the point of minimum loss from the modulator. Because of the slow variation of sinusoidal modulation, it is not very straightforward for generating ultrashort optical pulses (< 1ps) using this method.

For stable operation, the cavity length must precisely match the period of the modulation signal or some integer multiple of it. The most powerful technique to solve this is regenerative mode locking *i.e.*. a part of the output signal of the mode-locked laser is detected; the beatnote at the round-trip frequency is filtered out from the detector, and sent to an amplifier, which drives the loss modulator in the laser cavity. This procedure enforces synchronism if the cavity length undergoes fluctuations due to acoustic vibrations or thermal expansion. By using this method, highly stable mode-locked lasers have been achieved. The major advantage of active mode-locking is that it allows synchronized operation of the mode-locked laser to an external radio frequency (RF) source. This is very useful for optical fibre communication where synchronization is normally required between optical signal and electronic control signal. Also active mode-locked fibre can provide much higher repetition rate than passive mode-locking. Currently, fibre lasers and semiconductor diode lasers are the two most important types of lasers where active mode-locking are applied.

Dark soliton fibre lasers

In the non-mode locking regime,the first dark soliton fibre laser has been successfully achieved in an all-normal dispersion erbium-doped ber laser with a polarizer in cavity. Experimentally finding that apart from the bright pulse emission, under appropriate conditions the ber laser could also emit single or multiple dark pulses. Based on numerical simulations we interpret the dark pulse formation in the laser as a result of dark soliton shaping.

Multiwavelength fibre lasers

Recently,multiwavelength dissipative soliton in an all normal dispersion fibre laser passively mode-locked with a SESAM has been generated. It is found that depending on the cavity birefringence, stable single-, dual- and triple-wavelength dissipative soliton can be formed in the laser. Its generation mechanism can be traced back to the nature of dissipative soliton.

FIBRE DISK LASERS

Another type of fibre laser is the fibre disk laser. In such, the pump is not confined within the cladding of the fibre (as in the double-clad fibre), but pump light is delivered across the core multiple times because the core is coiled on itself like a rope. This configuration is suitable for power scaling in which many pump sources are used around the periphery of the coil.

FIBRE MANAGEMENT SYSTEM

A FMS (Fibre Management System) manages the fibre connections from outside of fibre rack to the fibre routers. Fibre cable duct containing many fibers come from far end sites and terminate on FMS using splicing technology. FMS has fibre in and fibre out ports. From fibre out port the fibre patch will go to fibre optics based router.

FIBRE OPTIC COUPLER

A fibre optic coupler is a device used in optical fibre systems with one or more input fibers and one or several output fibers. Light entering an input fibre can appear at one or more outputs and its power distribution potentially depending on the wavelength and polarization. Such couplers can be fabricated in different ways, for example by thermally fusing fibers so that their cores get into intimate contact. If all involved fibers are single-mode (supporting only a single mode per polarization direction for a given wavelength), there are certain physical restrictions on the performance of the coupler. In particular, it is not possible to combine two or more inputs of the same optical frequency into one single-polarization output without significant excess losses. However, such a restriction does not occur for different input wavelengths: there are couplers that can combine two inputs at different wavelengths into one output without exhibiting significant losses. Wavelength-sensitive couplers are used as multiplexers in wavelength-division multiplexing (WDM) telecom systems to combine several input channels with different wavelengths, or to separate channels.

FIBRE OPTIC FILTER

Fibre optic filter is an optical fibre instrument used for wavelength selection, which can select desired wavelengths to pass and reject the others. It is Widely used in DWDM systems dynamic wavelength selection, DWDM signal separation, optical performance monitoring, field tunable optical noise filtering and optical amplifier noise suppression, etc. Optical multiplexers (couplers) makes different wavelength coupling into an optical fibre and different wavelength carries different information. At the receiving end, if you want to separate desired wavelengths from optical fibre, it is necessary to use optical filter.

- Characteristics: Many wavelengths selection, channel spacing can be very small, low manufacturing cost, small polarization dependent loss, small design, fast installation, stable and reliable performance, different connector can be provided according to customer demand.
- Application: FTTX optical fibre communication systems, WDM system, LAN, CATV, fibre optic sensing, fibre amplifiers, fibre laser, measuring instruments, medical equipment.

PRINCIPLE

Grating spectrometer principle:

- The above drawing is a schematic diagram of grating separating the mixed lightwave. Mixed wavelengths•ië1•Aë2•Aë3•jfrom optical fibre, After the collimating lens (L1) post-directed to the grating•Cthe optical signals of different wavelengths due to the different diffraction angle, is focused to different position through the lens (L2),then the optical signal is coupled into different optical fibre for output. This is the principle of the grating spectrometer.
- Prism spectroscopic principle as shown in drawings, Its working principle is: Contains more than one optical signal wavelength of light,

As a result of Lens collimator, separated the light through the prism, After the separation of light through another lens is focused and coupled into the corresponding optical fibre in the spread. As well known, the propagation velocity of the different wavelengths in the same kind of material is not the same, That refractive index n (n = c / V) with wavelength dependent, If choose dn / dë material for prism, you can get large angular dispersion of skills and high colour resolution power.

- In addition, When the width of the prism surface is appropriately increased, and as far as possible to reduce the diameter of the collimating lens, You can get the best performance of the spectral effects, The lens, in the above system can be used to replace the self-focusing lens, its effect is completely the same.

Interference film filter principle:

- The interference film structure as shown in the drawings. The superposition of alternating dielectric film by both refractive index (n) of varying sizes. Its thickness is 1/4 wavelength, By different choices of the dielectric film constituting longwave pass, shortwave pass, and band pass filter. A high refractive index layer of the light reflected from its phase does not shift, The reflected light of the low refractive index layer whose phase is shifted by 180 degrees, Reflected several times through the interface of each film and the linear superposition of the transmitted light, when the optical wavelength of the optical path difference is equal to, or the same phase, multiple transmitted light will interfere with phase strengthened, to form a strong transmitted lightwaveinverting lightwave offset each other. Can be obtained a good filtering performance filters by appropriate design of the dielectric multilayer film systems.

OPTICAL FIBRE

- Optical fibre connector
- Fibre-optic communication
- Optical fibre cable
- Optical interconnect
- Fibre optic patch cord
- Telecommunications Industry Association
- Fibre
- Splitter ADSL

FIBERSCOPE

A Fiberscope is a flexible fibre optic bundle with an eyepiece on one end and a lens on the other that is used to examine and inspect small, hard to reach places such as the insides of machines, locks, and the human body. The concept of fibre optics began in 1854 when John Tyndall discovered that light could be conducted through a curved stream of water, proving that light could be bent. Then in 1930, Heinrich Lamm, a German medical student, became the first person to put together a bundle of optical fibers to carry an image. These discoveries led to the invention of endoscopes and fiberscopes. In the 1960s the endoscope was upgraded with glass fibre, a flexible material that allowed light to transmit, even when bent. While this provided users with the capability of real-time observation, it did not provide them with the ability to take photographs. In 1964 the fiberscope, the first gastro camera, was invented. It was the first time an endoscope had a camera that could take pictures. This innovation led to more careful observations, and more accurate diagnosis.

OPTICS

Fiberscopes work by utilizing the science of fibre optic bundles, which consist of numerous fibre optic cables. Fibre optic cables are made of optically pure glass and are as thin as a human's hair. The 3 main components to a fibre optic cable are:

- Core – The center made of high purity glass
- Cladding – The outer material surrounding the core that prevents light from leaking
- Buffer Coating – The protective plastic coating

There are two different fibre optic bundles in a Fiberscope, these include:

- Illumination Bundle – Designed to carry light from the source to the eyepiece.
- Imaging Bundle – Designed to carry an image from the lens to the eyepiece.

TOTAL INTERNAL REFLECTION

Fibre optic cables use total internal reflection to digitally carry information. When light travels from one medium to another it is refracted. If the light is traveling from a less dense medium to a dense medium it is refracted away from the normal. The opposite applies if the light is traveling from a dense medium to a less dense medium. In optic cables, light travels through the dense glass core (high refractive index) by constantly reflecting from the less dense cladding (lower refractive index). This happens because the surface of the core acts like a perfect mirror and the angle of the light is always larger than the critical angle.

COMPONENTS

- Eyepiece – Magnifies the image carried back by the imaging bundle so the human eye can view it.

- Imaging Bundle – Continuous strand of flexible glass fibers that transmit the image to the eyepiece.
- Distal Lens – The combination of micro lenses that take images and focus them into the small imaging bundle.
- Illumination System- A Fibre optic light guide that relays light from the source to the target area (eyepiece)
- Articulation System- The ability of the user to control the movement of the bending section of the fiberscope that is directly attached to the distal lens.
- Fiberscope Body – The control section that is designed to help aide one hand operation.
- Insertion Tube – Most of the length of the fiberscope, made to be durable and flexible. This protects the optical fibre bundle and the articulation cables.
- Bending Section – The most flexible part of the fiberscope, it connects the insertion tube to the distal viewing section.
- Distal Section – Where the ending points of both the illumination and imaging fibre bundle are.

MEDICAL APPLICATIONS

Fiberscopes are used in the medical field as a tool to help doctors and surgeons examine problems in a patient's body without having to make large incisions. This procedure is called an endoscopy. Doctors use this when they suspect that a patient's organ is infected, damaged, or cancerous. There are numerous types based on the area of the body being examined. They include:

- Arthroscopy – Joints
- Bronchoscopy – Lungs
- Colonoscopy – Colon
- Cystoscopy – Bladder
- Enteroscopy – Small Intestine
- Hysteroscopy – Uterus
- Laparoscopy – Abdomen/Pelvis
- Laryngoscopy – Larynx (voice box)
- Mediastinoscopy – Area between lungs
- Upper Gastrointestinal Endoscopy – Esophagus and upper intestinal tract

Although any medical technique has its potential risks, using a fiberscope for endoscopy has a very low risk of causing infection and blood loss.

OTHER APPLICATIONS

Although the fiberscope is mainly used for medical purposes, it is occasionally used for other purposes. These include

- Lock picking- Used to look inside the lock and see which pins need to be pushed to open the lock.
- Machine Inspection-Used to look inside machines to make sure everything is working properly.
- Computer Repair- Used to look at the small components inside a computer to figure out what is broken and how to fix it.
- Espionage- Used to look under doors or through small holes to spy on people.

FIGURE-8 LASER

A figure-8 laser is a fibre laser with a figure 8-shaped ring resonator. It is used for making pico- and femtosecond soliton pulses. The typical spectrum of such a laser consists of a wide central peak and a few narrow lateral peaks that are placed symmetrically around it. The amplitudes of the narrow peaks are the same as or less than that of the central peak.

Both loops of the resonator work as Sagnac loops. The active medium of the laser is optical fibre with its core doped with rare earth ions. It is placed asymmetrically with respect to the resonator loops to make a non-linear difference in phase between opposite waves, ensuring mode locking. In 1992 a figure-8 laser was built which had a smaller loop length of 1.6 m and a larger loop length of 60.8 m for generation of 315 fs pulses with interval 125 MHz.

FUSION SPLICING

Fusion splicing is the act of joining two optical fibers end-to-end using heat. The goal is to fuse the two fibers together in such a way that light passing through the fibers is not scattered or reflected back by the splice, and so that the splice and the region surrounding it are almost as strong as the virgin fibre itself. The source of heat is usually an electric arc, but can also be a laser, or a gas flame, or a tungsten filament through which current is passed.

PROCESS

The process of fusion splicing normally involves using localized heat to melt or fuse the ends of two optical fibers together. The splicing process begins by preparing each fibre end for fusion.

Stripping the fibre

Stripping is the act of removing the protective polymer coating around optical fibre in preparation for fusion splicing. The splicing process begins by preparing both fibre ends for fusion, which requires that all protective coating is removed or stripped from the ends of each fibre. Fibre optical stripping is usually carried out by a special stripping and preparation unit that uses hot sulphuric acid or a controlled flow of hot air to remove the coating. There are also mechanical tools used for stripping fibre which are similar to copper wire strippers. Fibre optical stripping and preparation equipment used in fusion splicing is commercially available through a small number of specialized companies, which usually also designs machines used for fibre optical recoating.

Cleaving the fibre

The fibre is then cleaved using the score-and-break method so that its endface is perfectly flat and perpendicular to the axis of the fibre. The quality of each fibre end is inspected using a microscope. In fusion splicing, splice loss is a direct function of the angles and quality of the two fibre-end faces. The closer to 90 degrees the cleave angle is the lower optical loss the splice will yield.

Splicing the fibers

Current fusion splicers are either core or cladding alignment. Using one of these methods the two cleaved fibers are automatically aligned by the fusion splicer in the x,y,z plane, then are fused together. Prior to removing the spliced fibre from the fusion splicer, a proof-test performed to ensure that the splice is strong enough to survive handling, packaging and extended use. The bare fibre area is protected either by recoating or with a splice protector. A splice protector is a heat shrinkable tube with a strength membrane.

A simplified optical splicing procedure includes:

1. Characteristics of placement of the splicing process.
2. Checking fibre optic splice closure content and supplementary kits.
3. Cable installation in oval outlet.
4. Cable preparation.
5. Organization of the fibers inside the tray.
6. Installing the heat shrinkable sleeve and testing it.

HARDWARE

The basic fusion splicing apparatus consists of two fixtures on which the fibers are mounted and two electrodes. These fixtures are often called sheath clamps. Inspection microscope assists in the placement of the prepared fibre ends into a fusion-splicing apparatus. The fibers are placed into the apparatus, aligned, and then fused together. Initially, fusion splicing used nichrome wire as the heating element to melt or fuse fibers together. New fusion-splicing techniques have replaced the nichrome wire with carbon dioxide (CO_2) lasers, electric arcs, or gas flames to heat the fibre ends, causing them to fuse together. The small size of the fusion splice and the development of automated fusion-splicing machines have made electric arc fusion (arc fusion) one of the most popular splicing techniques in commercial applications. Alternatives to fusion splicing include using optical fibre connectors or mechanical splices both of which have higher insertion losses, lower reliability and higher return losses than fusion splicing.

GLOW PLATE

Glow plates are sheets of glass or plastic that "glow" when light is supplied to one of their edges. The light source for a glow plate can be artificial, such as fluorescent light, or natural, with sunlight being directly exposed to the plate or fed through a fibre-optic system. A joint effort between Florida State University and Oak Ridge National Laboratory is focused on the design of a "spiral bio-reactor light sheet", which consists of a plexiglas sheet that has been micro-etched on one side and rolled into a spiral shape. Aside from aesthetic or utilitarian lighting purposes, much interest in using glow plates as a source of light comes from recent developments in algal cultivation.

GRADIENT-INDEX OPTICS

Gradient-index (GRIN) optics is the branch of optics covering optical effects produced by a gradual variation of the refractive index of a material. Such variations can be used to produce lenses with flat surfaces, or lenses that do not have the aberrations typical of traditional spherical lenses. Gradient-index lenses may have a refraction gradient that is spherical, axial, or radial.

The lens of the eye is the most obvious example of gradient-index optics in nature. In the human eye, the refractive index of the lens varies from approximately 1.406 in the central layers down to 1.386 in less dense layers of the lens (Hecht 1987, p. 178). This allows the eye to image with good resolution and low aberration at both short and long distances (Shirk *et al.*, 2006).

Another example of gradient index optics in nature is the common mirage of a pool of water appearing on a road on a hot day. The pool is actually an image of the sky, apparently located on the road since light rays are being refracted (bent) from their normal straight path. This is due to the variation of refractive index between the hot, less dense air at the surface of the road, and the denser cool air above it. The variation in temperature (and thus density) of the air causes a gradient in its refractive index, causing it to increase with height (Tsiboulia, 2003). This index gradient causes refraction of light rays (at a shallow angle to the road) from the sky, bending them into the eye of the viewer, with their apparent location being the road's surface.

The Earth's atmosphere acts as a GRIN lens, allowing observers to see the sun for a few minutes after it is actually below the horizon, and observers can also view stars that are below the horizon (Tsiboulia, 2003). This effect also allows for observation of electromagnetic signals from satellites after they have descended below the horizon, as in radio occultation measurements.

APPLICATIONS

The ability of GRIN lenses to have flat surfaces simplifies the mounting of the lens, which makes them useful where many very small lenses need to be mounted together, such as in photocopiers and scanners. The flat surface also allows a GRIN lens to be easily fused to an optical fibre, to produce collimated output.

In imaging applications, GRIN lenses are mainly used to reduce aberrations. The design of such lenses involves detailed calculations of aberrations as well as efficient manufacture of the lenses. A number of different materials have been used for GRIN lenses including optical glasses, plastics, germanium, zinc selenide, and sodium chloride.

Certain optical fibres (graded-index fibres) are made with a radially-varying refractive index profile; this design strongly reduces the modal dispersion of a multi-mode optical fibre. The radial variation in refractive index allows for a sinusoidal height distribution of rays within the fibre, preventing the rays from leaving the core. This differs from traditional optical fibres, which rely on total internal reflection, in that all modes of the GRIN fibres propagate at the same speed, allowing for a higher temporal bandwidth for the fibre (Moore, 1980).

MANUFACTURE

GRIN lenses are made by several techniques:

- Neutron irradiation (Sinai, 1971) – Boron-rich glass is bombarded with neutrons to cause a change in the boron concentration, and thus the refractive index of the lens.
- Chemical vapour deposition (Keck *et al.*, 1975) – Involving the deposition of different glass with varying refractive indexes, onto a surface to produce a cumulative refractive change.
- Partial polymerisation (Moore, 1973) – An organic monomer is partially polymerized using ultraviolet light at varying intensities to give a refractive gradient.
- Ion exchange (Hensler, 1975) – Glass is immersed into a liquid melt with lithium ions. As a result of diffusion, sodium ions in the glass are partially exchanged with lithium ones, with a larger amount of exchange occurring at the edge. Thus the sample obtains a gradient material structure and a corresponding gradient of the refractive index.
- Ion Stuffing (Mohr, 1979) – Phase separation of a specific glass causes pores to form, which can later be filled using a variety of salts or concentration of salts to give a varying gradient.

HISTORY

In 1854, J C Maxwell suggested a lens whose refractive index distribution would allow for every region of space to be sharply imaged. Known as the *Maxwell fisheye lens*, it involves a spherical index function and would be expected to be spherical in shape as well (Maxwell, 1854). This lens, however, is impractical to make and has little usefulness since, only points on the surface and within the lens are sharply imaged and extended objects suffer from extreme aberrations. In 1905, R W Wood used a dipping technique creating a gelatin cylinder with a refractive index gradient that varied symmetrically with the radial distance from the axis. Disk shaped slices of the cylinder were later shown to have plane faces with radial index distribution. He showed that even though the faces of the lens were flat, they acted like converging and diverging lens depending on whether the index was a decreasing or increasing relative to the radial distance (Wood, 1905). In 1964, was published a posthumous book of R. K. Luneburg where he described a lens that focuses incident parallel rays of light onto a point on the opposite surface of the lens (Luneburg, 1964). This also limits the applications of the lens, in that it is difficult to be used to focus visual light, however, it has some usefulness in microwave applications.

THEORY

An inhomogeneous gradient-index lens possesses a refractive index whose change follows the function $n = f(x, y, z)$ of the coordinates of the region of interest in the medium. According to Fermat's principle, the light path integral (L), taken along a ray of light joining any two points of a medium, is stationary relative to its value for any nearby curve joining the two points. The light path integral is given by the equation

$L = \int_{S_0}^{S} n\, ds$, where n is the refractive index and S is the arc length of the curve. If Cartesian coordinates are used, this equation is modified to incorporate the change in arc length for a spherical gradient, to each physical dimension:

$$L = \int_{S_0}^{S} n\,(x, y, z)\sqrt{x'^2 + y'^2 + z'^2}\, ds$$

where prime corresponds to d/ds (Marchand, 1978). The light path integral is able to characterize the path of light through the lens in a qualitative manner, such that the lens may be easily reproduced in the future.

The refractive index gradient of GRIN lenses can be mathematically modelled according to the method of production used. For example, GRIN lenses made from a radial gradient index material, such as SELFOC Microlens (Flores-Arias *et al.*, 2006), have a refractive index that varies according to:

$n_r = n_0 \left(1 - \frac{Ar^2}{2}\right)$, where n_r the refractive index at a distance, r, from the optical axis; n_0 is the design index on the optical axis, and A is a positive constant.

GUIDED RAY

A guided ray (also bound ray or trapped ray) is a ray of light in a multi-mode optical fibre, which is confined by the core. For step index fibre, light entering the fibre will be guided if it falls within the acceptance cone of the fibre, that is if it makes an angle with the fibre axis that is less than the acceptance angle,

$$\sin\theta \leq \sqrt{n_o^2 - n_c^2} ,$$

where

θ is the angle the ray makes with the fibre axis, *before* entering the fibre,

n_o is the refractive index along the central axis of the fibre, and

n_c is the refractive index of the cladding.

This result can be derived from Snell's law by considering the critical angle.

Rays that fall within this angular range are reflected from the core-cladding boundary by total internal reflection, and so are confined by the core. The confinement of light by the fibre can also be described in terms of bound modes or guided modes. This treatment is necessary when considering singlemode fibre, since the ray model does not accurately describe the propagation of light in this type of fibre.

FIBRE OPTIC GYROSCOPE

A fibre optic gyroscope (FOG) senses changes in orientation, thus performing the function of a mechanical gyroscope. However its principle of operation is instead based on the interference of light which has passed through a coil of optical fibre which can be as long as 5 km. The development of diode (semiconductor) lasers and low-loss single-mode optical fibre in the early 1970s for the telecommunications industry enabled Sagnac effect fibre optic gyros to be developed as practical devices.

OPERATION

Two beams from a laser are injected into the same fibre but in opposite directions. Due to the Sagnac effect, the beam travelling against the rotation experiences a slightly shorter path delay than the other beam. The resulting differential phase shift is measured through interferometry, thus translating one component of the angular velocity into a shift of the interference pattern which is measured photometrically.

Beam splitting optics launches light from a laser diode into two waves propagating in the clockwise and anticlockwise directions through a coil consisting of many turns of optical fibre.

The strength of the Sagnac effect is dependent on the *effective area* of the closed optical path: this is not simply the geometric area of the loop but is enhanced by the number of turns in the coil.

The FOG was first proposed by Vali and Shorthill in 1976. Development of both the passive interferometer type of FOG, or IFOG, and a newer concept, the passive ring resonator FOG, or RFOG, is proceeding in many companies and establishments worldwide.

ADVANTAGES

A FOG provides extremely precise rotational rate information, in part because of its lack of cross-axis sensitivity to vibration, acceleration, and shock. Unlike the classic spinning-mass gyroscope, the FOG has no moving parts and doesn't rely on inertial resistance to movement. Hence, this is perhaps the most reliable alternative to the mechanical gyroscope. Because of their intrinsic reliability, FOGs are used for high performance space applications.

The FOG typically shows a higher resolution than a ring laser gyroscope, but suffered from greater drift and worse scale factor performance until the end of the 1990s. FOGs are implemented in both open-loop and closed-loop configurations.

DISADVANTAGES

FOG requires calibration (determining which indication corresponds to zero angular velocity) while ring laser gyroscope does not (zero beat frequency always means zero angular velocity).

APPLICATIONS

1. FOGs are used in the inertial navigation systems of many guided missiles.
2. FOGs can be a navigation aid in remotely operated vehicles and autonomous underwater vehicles.
3. FOGs are used in surveying.

HALF ACCEPTANCE ANGLE

A ray of light traveling in an optical fibre must be incident at the critical angle on the internal surface of the fibre in order to have total internal reflection. When the ray is fed into the fibre, the refractive index of the traveling medium is different from what it is inside the optical fibre. So the angle of incidence should be such that after refraction, when the ray of light reaches the cladding, it does so at the critical angle.

GEORGE HOCKHAM

George Alfred Hockham FREng FIET (7 December 1938 – 16 September 2013) was a British engineer. He worked for over 40 years in theoretical analysis and design techniques applied to the solution of electromagnetic problems covering many different antenna types for radar, electronic warfare and communication systems. He coauthored the original paper on the application of cladded glass fibre as a transmission medium.

Born in Epsom, Surrey, Hockham received a BSc (Eng) degree from the Regent Street Polytechnic in 1961, and a PhD in 1969 from the University of London. He has worked for over 40 years in theoretical analysis and design techniques applied to the solution of electromagnetic problems covering many different antenna types for radar, electronic warfare and communication systems.

Hockham has also contributed significantly to the development of optical fibres for long distance communications systems. He proposed and published, together with a colleague Professor K. C. Kao, in 1966 the original paper on the application of the cladded glass fibre as the transmission medium – "Dielectric-fibre surface waveguides for optical frequencies" – for which he received the Rank Prize in Opto-Electronics in 1978. He is the holder of 16 scientific/technical patents and authored and co-authored 26 papers published in professional journals. He was previously Technical Director of Thorn EMI Electronics, Sensors Group, Technical Director in Plessey Radar, Director of Technology Plessey Electronics Systems and Manager of the Antenna and Microwave Laboratory, ITT Gilfillan, Los Angeles, Member of Advisory committees to the MoD and Academia. He is currently a Visiting Professor at Queen Mary, University of London.

FAMILY AND HOBBIES

George Alfred Hockham was born in 1938 in Epsom, Surrey, the only child of George and Elizabeth Hockham (née Elliott). His parents moved to Enfield when he was 6 months old and that is where he spent his childhood and early adulthood. He attended the local Albany Boys' Secondary School and later Regent Street Polytechnic where he obtained a BSc in Electrical Engineering. He started work at STL Harlow as a young graduate engineer in 1961.

He met his wife Mary in 1962 and they were married in 1964. Following his marriage he produced the most creative and productive work of his life. Two daughters and two sons were born in the next nine years. Cyril Connolly may well have deemed the pram in the hall the enemy of creativity but Hockham co-authored the paper that started it all in 1966. Fibre optics has changed the modern communications world.

Hockham was a keen swimmer in his youth and swam for his school and county. He later swam for STL Harlow in inter-companies competitions. He was also an enthusiastic member of the Regent Street Polytechnic Water Polo team during his undergraduate days – some would say too enthusiastic, spending more time in the water than at study. On giving up swimming he took up amateur motor cycle racing . He was a keen follower of Motor cycle racing and Formula 1 car racing.

AWARDS AND ACHIEVEMENTS

- Authored and co-authored 26 papers published in professional journals, including what is widely reported as the pioneering paper in fibre optics in 1966 the original paper – Dielectric Fibre Surface Waveguides for Optical Frequencies – with Charles Kao.

- PhD from University of London (1969)
- BSc(Eng) Regent Street Polytechnic (1961)
- Featured in BBC programme Tomorrow's World
- Rank Prize for Opto-Electronics 6 March 1978
- Visiting Professor Queen Mary, University of London
- Honorary Professor Beijing University of Posts and Telecommunications in the field of Microwave and Lightwave Communication (2008)
- Featured in Science Museum (London)
- Featured in British Genius Exhibition (1977)
- Fellow Institution of Electrical Engineers (1987)
- Fellow Royal Academy of Engineering (1995)
- Holder of 16 scientific / technical patents

WORK HISTORY

Hockham joined Standard Telecommunication Laboratories (STL) in Harlow following graduation in September 1961 working in the Microwave laboratory under Professor A E Karbowiak. The work here was on Trunk communication systems. During this period with the advent of the LASER there was now available an optical coherent source. By adopting optical frequencies that allowed much higher bandwidth for the transmission of information and the emphasis switched to one of finding a suitable optical waveguide. Several options were considered, the thin film waveguide proposed by Karbowiak which Hockham worked on but had the limitation that the electromagnetic field could not be contained laterally and proved to be of limited use. Another colleague investigated a confocal lens system. This comprised a periodic array of lenses displaced longitudinally where in theory the light beam was focused periodically. This too proved to be unacceptable as the whole structure had to be contained within a controlled environment *e.g.*. as soon as the temperature changed along the axis the beam was directed away from the axial direction and the light beam was lost- another failure.

At this time Professor Karbowiak left STL in 1964 and took up a position at the University of New South Wales in Sydney. Charles K. Kao was transferred into the group and we started looking at other options in particular the fibre.

Hockham started looking at the theoretical aspects in particular the loss due to discontinuities in the fibre and also the loss incurred when the fibre was curved, both were known to affect the performance and needed to be quantified as any one of these could have rendered the fibre approach unacceptable. The single fibre would need to be less than 1 micrometre in diameter to preserve single mode operation and also most of the energy is carried outside the fibe core to preserve the low loss, this too was a non-starter. However, if the core was surrounded by a cladding whose refractive index was close to that of the core a larger structure (in relative terms) could be accommodated. In this case most of the energy is now contained in the core and cladding regions of the fibre thus returning to the high losses. Charles Kao's part of the joint project was to investigate the losses in the glass material to determine if this could be reduced. All parts of the programme were successful leading now to a viable solution of a fibre optic communication system.

HYDROGEN DARKENING

Hydrogen darkening is a physical degradation of the optical properties of glass. Free hydrogen atoms are able to bind to the SiO_2 silica glass compound forming hydroxyl (OH) - a chemical compound that interferes with the passage of light through the glass. The problem is particularly relevant to fibre optic cables — particularly in oil and gas wells where fibre optic cables are used for distributed temperature sensing. Hydrogen can be present due to the cracking of hydrocarbons in the well. The darkening of the fibre can distort the DTS reading and possibly render the DTS system inoperable due to the optical loss budget being exceeded. To prevent this, coatings such as carbon are applied to the fibre, and hydrogen capturing gels are used to buffer the fibre, and other proprietary techniques may be used to prevent hydrogen atoms from reaching the glass fibre via the cable sheath.

HYDROXYL ION ABSORPTION

Hydroxyl ion absorption is the absorption in optical fibers of electromagnetic waves, including the near-infrared, due to the presence of trapped hydroxyl ions remaining from water as a contaminant. The hydroxyl (OH) ion, can penetrate glass during or after product fabrication, resulting in significant attenuation of discrete optical wavelengths, *e.g.*, centred at 1.383 ìm, used for communications via optical fibres.

INDEX-MATCHING MATERIAL

In optics, an index-matching material is a substance, usually a liquid, cement (adhesive), or gel, which has an index of refraction that closely approximates that of another object (such as a lens, material, fibre-optic, etc.). When two substances with the same index are next to each other, light passes from one to the other with neither reflection nor refraction. As such, they are used for various purposes in science, engineering, and art. For example, in a popular home experiment, a glass rod is made almost invisible by immersing it in an index-matched transparent fluid such as mineral spirits.

IN MICROSCOPY

In light microscopy, oil immersion is a technique used to increase the resolution of a microscope. This is achieved by immersing both the objective lens and the specimen in a transparent oil of high refractive index, thereby increasing the numerical aperture of the objective lens. Immersion oilsf**** are transparent oils that have specific optical and viscosity characteristics necessary for use in microscopy. Typical oils used have an index of refraction around 1.515. An oil immersion objective is an objective lens specially designed to be used in this way. The index of the oil is typically chosen to match the index of the microscope lens glass, and of the cover slip.

IN FIBRE OPTICS

In fibre optics and telecommunications, an index-matching material may be used in conjunction with pairs of mated connectors or with mechanical splices to reduce signal reflected in the guided mode (known as return loss) (see: Optical fibre connector). Without the use of an index-matching material, Fresnel reflections will occur at the smooth end faces of a fibre unless there is no fibre-air interface or other significant mismatch in refractive index. These reflections may be as high as -14 dB (*i.e.,* 14 dB below the optical power of the incident signal). When the reflected signal returns to the transmitting end, it may be reflected again and return to the receiving end at a level that is (28 plus twice the fibre loss) dB below the direct signal. The reflected signal will also be delayed by twice the delay time introduced by the fibre. The twice-reflected, delayed signal superimposed on the direct signal may noticeably degrade an analog baseband intensity-modulated video signal. Conversely, for digital transmission, the reflected signal will often have no practical effect on the detected signal seen at the decision point of the digital optical receiver except in marginal cases where bit-error ratio is significant. However, certain digital transmitters such as those employing a Distributed Feedback Laser may be affected by back reflection and then fall outside specifications such as Side Mode Suppression Ratio, potentially degrading system bit error ratio, so networking standards intended for DFB lasers may specify a back-reflection tolerance such as -10dB for transmitters so that they remain within specification even without index matching. This back-reflection tolerance might be achieved using an optical isolator or by way of reduced coupling efficiency.

For some applications, instead of standard polished connectors (*e.g.*. FC/PC), angle polished connectors (*e.g.*. FC/APC) may be used, whereby the non-perpendicular polish angle greatly reduces the ratio of reflected signal launched into the guided mode even in the case of a fibre-air interface.

IN ART CONSERVATION

If a sculpture is broken into several pieces, art conservators may reattach the pieces using an adhesive such as Paraloid B-72 or epoxy. If the sculpture is made of a transparent or semitransparent material (such as glass), the seam where the pieces are attached will usually be much less noticeable if the refractive index of the adhesive matches the refractive index of the surrounding object. Therefore, art conservators may measure the index of objects and then use an index-matched adhesive. Similarly, losses (missing sections) in transparent or semitransparent objects are often filled using an index-matched material.

IN OPTICAL COMPONENT ADHESIVES

Certain optical components, such as a Wollaston prism or Nicol prism, are made of multiple transparent pieces that are directly attached to each other. The adhesive is usually index-matched to the pieces. Historically, Canada balsam was used in this application, but it is now more common to use epoxy or other synthetic adhesives.

INTERCONNECT BOTTLENECK

The interconnect bottleneck refers to limits on integrated circuit (IC) performance due to connections between components instead of their internal speed. In 2006 it was predicted to be a "looming crisis" by 2010.

Improved performance of computer systems has been achieved, in large part, by downscaling the IC minimum feature size. This allows the basic IC building block, the transistor, to operate at a higher frequency, performing more computations per second. However, downscaling of the minimum feature size also results in tighter packing of the wires on a microprocessor, which increases parasitic capacitance and signal propagation delay. Consequently, the delay due to the communication between the parts of a chip becomes comparable to the computation delay itself. This phenomenon, known as an "interconnect bottleneck", is becoming a major problem in high-performance computer systems.

This interconnect bottleneck can be solved by utilizing optical interconnects to replace the long metallic interconnects. Such hybrid optical/electronic interconnects promises better performance even with larger designs. Optics has widespread use in long-distance communications; still it has not yet been widely used in chip-to-chip or on-chip interconnections because they (in centimeter or micrometer range) are not yet industry-manufacturable owing to costlier technology and lack of fully mature technologies. As optical interconnections move from computer network applications to chip level interconnections, new requirements for high connection density and alignment reliability have become as critical for the effective utilization of these links. There are still many materials, fabrication, and packaging challenges in integrating optic and electronic technologies.

LEAKY MODE

A leaky mode or tunneling mode in an optical fibre or other waveguide is a mode having an electric field that decays monotonically for a finite distance in the transverse direction but becomes oscillatory everywhere beyond that finite distance. Such a mode gradually "leaks" out of the waveguide as it travels down it, producing attenuation even if the waveguide is perfect in every respect. In order for a leaky mode to be definable as a mode, the relative amplitude of the oscillatory part (the leakage rate) must be sufficiently small that the mode substantially maintains its shape as it decays.

Leaky modes correspond to leaky rays in the terminology of geometric optics. The propagation of light through optical fibre can take place via meridional rays or skew rays. These skew rays suffer only partial reflection while meridional rays are completely guided. Thus the modes allowing propagation of skew rays are called leaky modes. Some optical power is lost into clad due to these modes.

MANAKOV SYSTEM

Maxwell's Equations, when converted to cylindrical coordinates, and with the boundary conditions for an optical fibre while including birefringence as an effect taken into account, will yield the coupled non-linear Schrödinger equations. After employing the Inverse scattering transform (a procedure analogous to the Fourier Transform and Laplace Transform) on the resulting equations, the Manakov system is then obtained. The most general form of the Manakov system is as follows:

$$v_1' = -i\,\xi\,v_1 + q_1\,v_2 + q_2 v_3$$

$$v_2' = -q_1^*\,v_1 + i\,\xi\,v_2$$

$$v_3' = -q_2^*\,v_1 + i\,\xi\,v_3\,.$$

It is a coupled system of linear ordinary differential equations. The functions q_1, q_2 represent the envelope of the electromagnetic field as an initial condition.

For theoretical purposes, the integral equation version is often very useful. It is as follows:

$$\lim_{x\to a} e^{i\xi x} v_1 - \lim_{x\to b} e^{i\xi x} v_1 = \int_a^b [e^{i\xi x} q_1\, v_2 + e^{i\xi x} q_2\, v_3]dx$$

$$\lim_{x\to a} e^{-i\xi x} v_2 - \lim_{x\to b} e^{i\xi x} v_2 = -\int_a^b e^{-i\xi x} q_1^*\, v_1\, dx$$

$$\lim_{x\to a} e^{-i\xi x} v_3 - \lim_{x\to b} e^{i\xi x} v_3 = -\int_a^b e^{-i\xi x} q_2^*\, v_1\, dx$$

One may make further substitutions and simplifications, depending on the limits used and the assumptions about boundary or initial conditions. One important concept is that ξ is complex; assumptions must be made about this eigenvalue parameter. If a non-zero solution is desired, the imaginary part of the eigenvalue cannot change sign; accordingly, most researchers take the imaginary part to be positive.

MANDREL WRAPPING

In multimode fibre optics, mandrel wrapping is a technique used to preferentially attenuate high-order mode power of a propagating optical signal. Consequently, if the fibre is propagating substantial energy in affected modes, the modal distribution will be changed.

A cylindrical rod wrap consists of a specified number turns of fibre on a mandrel of specified size, depending on the fibre characteristics and the desired modal distribution. It has application in optical transmission performance tests, to create a defined mode power distribution or to prevent multimode propagation in single mode fibre. If the launch fibre is fully filled ahead of the mandrel wrap, the higher-order modes will be stripped off, leaving only lower-order modes. If the launch fibre is underfilled, for example as a consequence of being energized by a laser diode or edge-emitting LED, there will be no effect on the mode power distribution or loss measurements.

In multimode fibre, mandrel wrapping is used to eliminate the effect of "transient loss", the tendency of high order modes to experience higher loss than lower order modes. Numerical addition (in decibels) of the measured loss of multiple fibre segments and/or components overestimates the loss of the concatenated set if each segment or component has been measured with a full mode power distribution.

In single mode optical fibre measurements, it is used to enforce true single mode propagation at wavelengths near or below the theoretical cutoff wavelength, at which substantial power can exist in a higher order mode group. In this use, it is commonly termed a High Order Mode Filter (HOMF).

Ultimately, the effect of mandrel wrapping on optical measurements depends on the propagating mode power distribution. An additional loss mechanism has no effect unless power is present in the affected modes.

PRINCIPLE OF OPERATION

The effect of physically bending an optical fibre around a cylindrical form is to slightly modify the effective refractive index in the curved region, which locally reduces the effective mode volume of the fibre. This causes optical power in the highest order modes to become unguided, or so weakly guided as to be released into an unbound state, absorbed by the fibre coating or completely ejected from the fibre. The practical effect of mandrel wrapping is to attenuate optical power propagating in the highest order modes. Lower order modes are unaffected, experiencing neither increased loss nor conversion into other modes (mode mixing).

DETERMINATION OF APPROPRIATE MANDREL WRAP CONDITIONS

The mandrel diameter and number of turns are chosen to eliminate certain modes in a reproducible way. It is empirically observed that more than 5 full 360 degree wraps creates little additional loss, so 3 to 5 turns are commonly specified.

The mandrel diameter affects how far into the mode volume the modal unbinding occurs. Experimentally, one plots the transmitted power from a wrapped fibre into which a uniform modal power distribution has been excited, as a function of mandrel diameter, maintaining a constant number of turns. This reveals step-like reductions in transmitted power as the diameter decreases, where each step is the point at which the mandrel is beginning to affect the next-lower mode group. For best measurement reproducibility, one would select a diameter that is not near such a transition, although this may not be possible if measurements must be performed over a range of wavelengths. Total mode volume in a fibre is a function of wavelength, so the mandrel diameter at which the mode group transitions occur will change with wavelength.

MATERIAL DISPERSION COEFFICIENT

In an optical fibre, the material dispersion coefficient, $M(\lambda)$, characterizes the amount of pulse broadening by material dispersion per unit length of fibre and per unit of spectral width. It is usually expressed in picoseconds per (nanometre·kilometre). For many optical fibre materials, $M(\lambda)$ approaches zero at a specific wavelength λ_0 between 1.3 and 1.5 ìm. At wavelengths shorter than λ_0, $M(\lambda)$ is negative and increases with wavelength; at wavelengths longer than λ_0, $M(\lambda)$ is positive and decreases with wavelength. Pulse broadening caused by material dispersion in a unit length of optical fibre is given by the product of $M(\lambda)$ and spectral width ($\Delta\lambda$).

MICRODUCTS

Microducts are small ducts for the installation of small microduct fibre optic cables. They have a size ranging from typically 3 to 16 mm and are installed as bundles in larger ducts.

Microducts are typically small-diameter, flexible, or semi-flexible ducts designed to provide clean, continuous, low-friction paths for placing optical cables that have relatively low pulling tension limits. As stated in industry requirements document Telcordia GR-3155, *Generic Requirements for Microducts for Fibre Optic Cables,* microduct products are expected to:

- Be compatible with existing construction designs and building configurations for both riser- and plenum-rated applications, including cable blowing apparatus.
- Allow cables to be safely deployed through pull lines or strings using less than 50 lbs of force, and through cable blowing techniques at typical deployment speeds of 100-200 feet per minute.

TYPES OF DUCTS

GR-3155 states that the basic types of duct are smoothwall, corrugated, and ribbed. The selection of a particular duct design is dependent on those characteristics that are important to the end-user. The need for a specific characteristic or combination of characteristics such as pulling strength, flexibility, or the lowest coefficient of friction will dictate the type of duct required.

DUCT OPTIONS

Ducts can be purchased with a variety of options or features. One such enhancement is pre-lubrication. Pre-lubricated ducts may be either permanently impregnated with anti-friction compounds or coated with liquid lubricant during manufacture (see GR-3155). This may or may not eliminate the need for supplementary lubrication when pulling cable into the duct. Before using a supplementary lubricant with a pre-lubricated duct, the user should check with the manufacturer to determine if the added lubricant is compatible with the pre-lubricated surface of the duct. Failure to do this may result in the cable seizing up rather than reducing the friction coefficient of the duct.

PLACEMENT OF CABLE

As indicated in GR-3155, cable is typically placed into the duct in one of three ways:

- It may be pre-installed by the duct manufacturer during the extrusion process.
- It may be pulled into the duct using a mechanically assisted or hand-drawn pull line.
- It may be blown into the duct using a high air volume cable blowing apparatus.

Pre-installed cable

When cable is pre-installed, the duct manufacturer extrudes the duct directly over the optical cable. Tight control of the duct temperature during the manufacturing process is essential to ensure that the duct does not stick to the cable as it cools. At the completion of the process, all of the fibers in the optical cable must be tested to ensure that no damage has occurred.

Cable pulling

A common cable installation technique for fibre cables remains cable pulling. After the duct is placed, a high-strength pull line is blown into the duct (if one has not already been pre-installed by the duct manufacturer). The pull line is attached to one end of the cable and is used to pull the cable through the duct.

Cable blowing

Traditional cable pulling methods are very sensitive to the condition of the duct and to the number of bends and undulations throughout the duct route. Therefore, for microducts, Air-Blown (AB) cable installation techniques are expected to be the most useful. AB cable installation requires the use of a device that injects a high volume of air into the duct, at pressures as high as 20-25 psi. The viscous drag forces generated by the rushing air along the length of the cable act to reduce or overcome the friction between the cable and the duct.

TRADITIONAL DUCT INSTALLATION

For telecommunications, cables can be installed in water, in air or underground. In the latter case, the cables might be direct buried or installed in ducts. The first is more common for copper balanced cables; the latter for fibre optic cables. The ducts in which the fibre optic cables are installed are usually made of polyethylene. They have a size ranging from typically 25 mm to 100 mm. Sometimes they are installed as subducts in larger ducts. These larger ducts can also consist of other materials, like concrete. The installation of fibre optic cables in ducts can be done by pulling or by cable jetting.

PROBLEMS WITH FIBRE OPTIC CABLES

It is more difficult to make branching fibre optic networks in the access network than it is for copper balanced cables. Splicing optical fibres is much more difficult than connecting copper wires. In Fibre to the Home (FTTH), where a lot of branches are present in the network, an Optical Distribution Network is used to branch the cables from a roadside cabinet or pit that contains optical equipment and is fed from the Central Office

MICRODUCT CABLING

With microduct cabling, bundles of small microducts are installed in larger protective ducts. This can be done by jetting for example. Bundles of microducts can also be factory pre-installed. The microducts can be branched very easily in the network. At any place of choice, a window cut is made in the protective duct and the microduct of choice is cut. This microduct is then connected, using a simple push/pull connector, to a microduct that branches to the desired location. After all connections are made, an individual microduct path has been created in the network. A microduct cable can then be jetted through the microduct, without the need to make a splice.

ADVANTAGES OVER TRADITIONAL CABLING

1. A branch can be made simply, any place, at any time.
2. Low initial costs.
3. The network can grow on demand.

4. Easy to install microduct routes in occupied ducts.
5. Easy to replace old cables through the network.
6. Possibility to migrate from copper balanced cables to fibre optic cables.

PRACTICE

Today the microduct cabling technology is used more and more, all over the world. The fibre counts have grown up to 96 per cable and can be installed in microducts of only 8 mm inner diameter. Bundles of microducts can be jetted over 1500 m or more. Microduct cables can even be jetted over 3.5 km in one single shot. More length without splice is reached by placing jetting equipment in tandem.

MODAL DISPERSION

Modal dispersion is a distortion mechanism occurring in multimode fibers and other waveguides, in which the signal is spread in time because the propagation velocity of the optical signal is not the same for all modes. Other names for this phenomenon include multimode distortion, multimode dispersion, modal distortion, intermodal distortion, intermodal dispersion, and intermodal delay distortion.

In the ray optics analogy, modal dispersion in a step-index optical fibre may be compared to multipath propagation of a radio signal. Rays of light enter the fibre with different angles to the fibre axis, up to the fiber's acceptance angle. Rays that enter with a shallower angle travel by a more direct path, and arrive sooner than rays that enter at a steeper angle (which reflect many more times off the boundaries of the core as they travel the length of the fibre). The arrival of different components of the signal at different times distorts the shape.

Modal dispersion limits the bandwidth of multimode fibers. For example, a typical step-index fibre with a 50 μm core would be limited to approximately 20 MHz for a one kilometer length, in other words, a bandwidth of 20 MHz·km. Modal dispersion may be considerably reduced, but never completely eliminated, by the use of a core having a graded refractive index profile. However, multimode graded-index fibers having bandwidths exceeding 3.5 GHz·km at 850 nm are now commonly manufactured for use in 10 Gbit/s data links.

Modal dispersion should not be confused with chromatic dispersion, a distortion that results due to the differences in propagation velocity of different wavelengths of light. Modal dispersion occurs even with an ideal, monochromatic light source.

A special case of modal dispersion is polarization mode dispersion (PMD), a fibre dispersion phenomenon usually associated with single-mode fibers. PMD results when two modes that normally travel at the same speed due to fibre core geometric and stress symmetry (for example, two orthogonal polarizations in a waveguide of circular or square cross-section), travel at different speeds due to random imperfections that break the symmetry.

TROUBLESHOOTING

In multimode optical fibre with many wavelengths propagating, it is sometimes hard to identify the dispersed wavelength out of all the wavelengths that are present, if there is not yet a service degradation issue. One can compare the present optical power of each wavelength to the designed values and look for differences. After that, the optical fibre is tested end to end. If no loss is found, then most probably there is dispersion with that particular wavelength. Normally engineers start testing the fibre section by section until they reach the affected section; all wavelengths are tested and the affected wavelength produces a loss at the far end of the fibre. One can easily calculate how much of the fibre is affected and replace that part of fibre with a new one. Replacement of optical fibre is only required when there is an intense dispersion and service is being affected; otherwise various methods can be used to compensate for the dispersion.

MODE FIELD DIAMETER

In fibre optics, the mode field diameter (MFD) is an expression of distribution of the irradiance, *i.e.*, the optical power per unit area, across the end face of a single-mode fibre. For a Gaussian intensity (*i.e.*, power density, W/m) distribution in a single-mode optical fibre, the mode field diameter is that at which the electric and magnetic field strengths are reduced to 1/e of their maximum values, *i.e.*, the diameter at which power density is reduced to 1/e of the maximum power density, because the power density is proportional to the square of the field strength.

MODE SCRAMBLER

In telecommunications, a mode scrambler mode mixer is a device for inducing mode coupling in an optical fibre, or a device that, itself, exhibits a uniform output intensity profile independent of the input mode volume or modal excitation condition. Mode scramblers are used to provide a modal distribution that is independent of the optical source for purposes of laboratory, manufacturing, or field measurements or tests. Mode scramblers are primarily used to improve reproducibility of multimode fibre bandwidth measurements.

If multimode fibre bandwidth is measured using a laser diode directly coupled to its input, the resulting measurement can vary by as much as an order of magnitude. This measurement variability is due to the combination of differences in laser output characteristics (emitted mode power distribution) and the differential mode delay of the fibre. Differential mode delay is the difference in the time delays amongst the fiber's propagating modes caused by imperfections or non-ideality of the fibre refractive index profile.

The primary purpose of a mode scrambler is to create a uniform, overfilled launch condition that can be easily reproduced on multiple measurement systems, so that measurement systems have essentially the same launch conditions and can measure approximately the same bandwidth despite having different laser sources. These were used for this purpose in the first U.S. NIST round-robins on multimode fibre. The overfilled launch (OFL) was created to reduce measurement variability, and improve concatenation estimates for multimode fibers, used at that time for telecom 'long haul' (*e.g.*., 7-10 km 850 nm or 20-30 km 1300 nm) systems.

When the telecom industry converted to near-exclusive use of single-mode fibre ca. 1984, multimode fibre was re-purposed for use in LANs, such as Fibre Distributed Date Interface (FDDI), then under development. The ouptut modal power distribution of a mode scrambler is similar to the surface-emitters used in those first LAN transmitters, but this was fortuitous coincidence. On average, but not in every case, the OFL bandwidth measured using a mode scrambler is lower than that produced by excitation of a partial mode volume (restricted mode launch or RML), such as occurs with directly coupled laser diodes.

There are two common types of mode scramblers: the "Step-Graded-Step" (S-G-S) and the "Step Index with Bends". The S-G-S mode scrambler is actually an assembly, a fusion-spliced concatenation of a step-index profile, a graded-index profile and another step-index profile fibre. Typically, each segment is approximately 1 meter long, and may use segments of unconventional size to produce the distribution required according to core size of fibre to be tested. Unconventional fibre size was not an issue, as they were developed by fibre manufacturers, but some test equipment has difficulty complying with revised qualification standards, and now use "Step Index with Bends" mode scramblers, which can be adjusted to purpose. Step Index with Bend mode scramblers are created simply by routing a specially designed step-index multimode fibre through a series of small radius bends, or by compressing fibre against surfaces with specific roughness. The implementations are simple, but generally less reproducibile, and require care to avoid over-stressing the fibre.

A mode scrambler can be characterized and qualified by measuring its near field and far field distributions, as well as by measuring one of these distributions while restricting the other. Guidelines for constructing a mode scrambler and qualifying its output can be found in the ANSI/TIA/EIA-455-54 fibre optic test procedure (FOTP).

MODE VOLUME

In fibre optics, mode volume is the number of bound modes that an optical fibre is capable of supporting.

The mode volume M is approximately given by $\frac{4V^2}{\pi^2}$ and $\frac{V^2}{2}\left(\frac{g}{g+2}\right)$, respectively for step-index and power-law index profile fibers, where g is the profile parameter, and V is the normalized frequency, *which must be greater than 5 for this approximation to be valid.*

NORMALIZED FREQUENCY (FIBRE OPTICS)

In an optical fibre, the normalized frequency, V (also called the V number), is given by

$$V = \frac{2\pi a}{\lambda}\sqrt{n_1^2 - n_2^2} = \frac{2\pi a}{\lambda} NA$$

where a is the core radius, λ is the wavelength in vacuum, n_1 is the maximum refractive index of the core, n_2 is the refractive index of the homogeneous cladding, and applying the usual definition of the numerical aperture NA.

In multimode operation of an optical fibre having a power-law refractive index profile, the approximate number of bound modes (the mode volume), is given by

$$\frac{V^2}{2}\left(\frac{g}{g+2}\right)$$

where g is the profile parameter, and V is the normalized frequency, which must be greater than 5 for the approximation to be valid.

For a step index fibre, the mode volume is given by $4V/ð$. For single-mode operation is required that $V < 2.4048$, which is the first root of the Bessel function J_0.

NUMERICAL APERTURE

In optics, the numerical aperture (NA) of an optical system is a dimensionless number that characterizes the range of angles over which the system can accept or emit light. By incorporating index of refraction in its definition, NA has the property that it is constant for a beam as it goes from one material to another provided there is no optical power at the interface. The exact definition of the term varies slightly between different areas of optics. Numerical aperture is commonly used in microscopy to describe the acceptance cone of an objective (and hence its light-gathering ability and resolution), and in fibre optics, in which it describes the range of angles within which light that is incident on the fibre will be transmitted along it.

GENERAL OPTICS

In most areas of optics, and especially in microscopy, the numerical aperture of an optical system such as an objective lens is defined by

$$NA = n \sin\theta$$

where n is the index of refraction of the medium in which the lens is working, and *è* is the half-angle of the maximum cone of light that can enter or exit the lens. In general, this is the angle of the real marginal ray in the system. Because the index of refraction is included, the NA of a pencil of rays is an invariant as a pencil of rays passes from one material to another through a flat surface. This is easily shown by rearranging Snell's law to find that $n \sin\theta$ is constant across an interface.

In air, the angular aperture of the lens is approximately twice this value (within the paraxial approximation). The NA is generally measured with respect to a particular object or image point and will vary as that point is moved. In microscopy, NA generally refers to object-space NA unless otherwise noted.

In microscopy, NA is important because it indicates the resolving power of a lens. The size of the finest detail that can be resolved is proportional to λ/2NA, where λ is the wavelength of the light. A lens with a larger numerical aperture will be able to visualize finer details than a lens with a smaller numerical aperture. Assuming quality (diffraction limited) optics, lenses with larger numerical apertures collect more light and will generally provide a brighter image, but will provide shallower depth of field. Numerical aperture is used to define the "pit size" in optical disc formats.

Numerical aperture versus f-number

Numerical aperture is not typically used in photography. Instead, the angular aperture of a lens (or an imaging mirror) is expressed by the f-number, written *f*/# or *N*, which is defined as the ratio of the focal length *f* to the diameter of the entrance pupil *D*:

$$N = f / D$$

This ratio is related to the image-space numerical aperture when the lens is focused at infinity. Based on the diagram at the right, the image-space numerical aperture of the lens is:

$$NA_i = n\sin\theta = n\sin\left[\arctan\left(\frac{D}{2f}\right)\right] \approx n\frac{D}{2f}$$

thus $N \approx \frac{1}{2NA_i}$, assuming normal use in air ($n = 1$).

The approximation holds when the numerical aperture is small, but it turns out that for well-corrected optical systems such as camera lenses, a more detailed analysis shows that N is almost exactly equal to $1/(2NA_i)$ even at large numerical apertures. As Rudolf Kingslake explains, "It is a common error to suppose that the ratio $[D/2f]$ is actually equal to $\tan\theta$, and not $\sin\theta$... The tangent would, of course, be correct if the principal planes were really plane. However, the complete theory of the Abbe sine condition shows that if a lens is corrected for coma and spherical aberration, as all good photographic objectives must be, the second principal plane becomes a portion of a sphere of radius *f* centered about the focal point, ..." In this sense, the traditional thin-lens definition and illustration of f-number is misleading, and defining it in terms of numerical aperture may be more meaningful.

Working (effective) f-number

The f-number describes the light-gathering ability of the lens in the case where the marginal rays on the object side are parallel to the axis of the lens. This case is commonly encountered in photography, where objects being photographed are often far from the camera. When the object is not distant from the lens, however, the image is no longer formed in the lens's focal plane, and the f-number no longer accurately describes the light-gathering ability of the lens or the image-side numerical aperture. In this case, the numerical aperture is related to what is sometimes called the "working f-number" or "effective f-number." A practical example of this is, that when focusing closer, with *e.g.*. a macro lens, the lens' effective aperture becomes smaller, from *e.g.*. f/22 to f/45, thus affecting the exposure.

The working f-number is defined by modifying the relation above, taking into account the magnification from object to image:

$$\frac{1}{2NA_i} = N_w = (1-m)N,$$

where N_w is the working f-number, m is the lens's magnification for an object a particular distance away, and the NA is defined in terms of the angle of the marginal ray as before. The magnification here is typically negative; in photography, the factor is sometimes written as $1 + m$, where m represents the absolute value of the magnification; in either case, the correction factor is 1 or greater. The two equalities in the equation above are each taken by various authors as the definition of working f-number, as the cited sources illustrate. They are not necessarily both exact, but are often treated as if they are. The actual situation is more complicated — as Allen R. Greenleaf explains, "Illuminance varies inversely as the square of the distance between the exit pupil of the lens and the position of the plate or film. Because the position of the exit pupil usually is unknown to the user of a lens, the rear conjugate focal distance is used instead; the resultant theoretical error so introduced is insignificant with most types of photographic lenses."

Conversely, the object-side numerical aperture is related to the f-number by way of the magnification (tending to zero for a distant object):

$$\frac{1}{2NA_o} = \frac{m-1}{m}N$$

LASER PHYSICS

In laser physics, the numerical aperture is defined slightly differently. Laser beams spread out as they propagate, but slowly. Far away from the narrowest part of the beam, the spread is roughly linear with distance—the laser beam forms a cone of light in the "far field". The relation used to define the NA of the laser beam is the same as that used for an optical system,

$$NA = n \sin\theta,$$

but è is defined differently. Laser beams typically do not have sharp edges like the cone of light that passes through the aperture of a lens does. Instead, the irradiance falls off gradually away from the center of the beam. It is very common for the beam to have a Gaussian profile. Laser physicists typically choose to make è the *divergence* of the beam: the far-field angle between the propagation direction and the distance from the beam axis for which the irradiance drops to 1/e times the wavefront total irradiance. The NA of a Gaussian laser beam is then related to its minimum spot size by

$$NA \simeq \frac{\lambda_0}{\pi w_0},$$

where λ_0 is the vacuum wavelength of the light, and $2w_0$ is the diameter of the beam at its narrowest spot, measured between the 1/e irradiance points ("Full width at e maximum of the intensity"). This means that a laser beam that is focused to a small spot will spread out quickly as it moves away from the focus, while a large-diameter laser beam can stay roughly the same size over a very long distance.

FIBRE OPTICS

A multi-mode optical fibre will only propagate light that enters the fibre within a certain cone, known as the acceptance cone of the fibre. The half-angle of this cone is called the acceptance angle, θ_{max}. For step-index multimode fibre, the acceptance angle is determined only by the indices of refraction of the core and the cladding:

$$n \sin\theta_{max} = \sqrt{n_{core}^2 - n_{clad,}^2}$$

where n_{core} is the refractive index of the fibre core, and n_{clad} is the refractive index of the cladding. While the core will accept light at higher angles, those rays will not totally reflect off the core–cladding interface, and so will not be transmitted to the other end of the fibre.

When a light ray is incident from a medium of refractive index n to the core of index n_{core} at the maximum acceptance angle, Snell's law at the medium–core interface gives

$$n \sin \theta_{\max} = n_{\text{core}} \sin \theta_r .$$

From the geometry of the above figure we have:

$$\sin \theta_r = \sin (90° - \theta_c) = \cos \theta_c$$

where $\theta_c = \sin^{-1} \frac{n_{\text{clad}}}{n_{\text{core}}}$ is the critical angle for total internal reflection.

Substituting cos θ_c for sin θ_r in Snell's law we get:

$$\frac{n}{n_{\text{core}}} \sin \theta_{\max} = \cos \theta_c .$$

By squaring both sides

$$\frac{n^2}{n_{\text{core}}^2} \sin \theta_{\max} = \cos \theta_c = 1 - \sin^2 \theta_c = 1 - \frac{n_{\text{clad}}^2}{n_{\text{core}}^2} .$$

Solving, we find the formula stated above:

$$n \sin \theta_{\max} = \sqrt{n_{\text{core}}^2 - n_{\text{clad}}^2} ,$$

This has the same form as the numerical aperture in other optical systems, so it has become common to *define* the NA of any type of fibre to be

$$NA = \sqrt{n_{\text{core}}^2 - n_{\text{clad}}^2} ,$$

where n_{core} is the refractive index along the central axis of the fibre. Note that when this definition is used, the connection between the NA and the acceptance angle of the fibre becomes only an approximation. In particular, manufacturers often quote "NA" for single-mode fibre based on this formula, even though the acceptance angle for single-mode fibre is quite different and cannot be determined from the indices of refraction alone. The number of bound modes, the mode volume, is related to the normalized frequency and thus to the NA. In multimode fibers, the term *equilibrium numerical aperture* is sometimes used. This refers to the numerical aperture with respect to the extreme exit angle of a ray emerging from a fibre in which equilibrium mode distribution has been established.

OPEN FIBRE CONTROL

In telecommunication, Open fibre control is a protocol to ensure that both ends of a fibre optic cable are connected before laser signals are transmitted in order to protect people from eye damage. When a device is turned on, it sends out low powered light. If it does not receive light back, it assumes that the fibre is not connected. When it receives light, it assumes that both ends of the fibre are connected and it switches the laser to full power. If one of the devices stops receiving light, it will revert to the low power mode.

OPTICAL POWER METER

An optical power meter (OPM) is a device used to measure the power in an optical signal. The term usually refers to a device for testing average power in fibre optic systems. Other general purpose light power measuring devices are usually called radiometers, photometers, laser power meters, light meters or lux meters.

A typical optical power meter consists of a calibrated sensor, measuring amplifier and display. The sensor primarily consists of a photodiode selected for the appropriate range of wavelengths and power levels. On the display unit, the measured optical power and set wavelength is displayed. Power meters are calibrated using a traceable calibration standard such as a NIST standard.

A traditional optical power meter responds to a broad spectrum of light, however the calibration is wavelength dependent. This is not normally an issue, since the test wavelength is usually known, however it has a couple of drawbacks. Firstly, the user must set the meter to the correct test wavelength, and secondly if there are other spurious wavelengths present, then wrong readings will result.

Sometimes optical power meters are combined with a different test function such as an Optical Light Source (OLS) or Visual Fault Locator (VFL), or may be a sub-system in a much larger instrument. When combined with a light source, the instrument is usually called an Optical Loss Test Set.

Optical Loss Test Sets (OLTS) are available in dedicated hand held instruments and platform-based modules to suit various network architectures and test requirements. They are used to measure optical power and power loss, and reflectance and reflected power loss. The products may also be used as optical sources or optical power meters, or to measure optical return loss or event reflectance.

Three types of equipment can be used to measure optical power loss:

1. *Component equipment* - Optical Power Meters (OPMs) and Stabilized Light Sources (SLSs) are packaged separately, but when used together they can provide a measurement of end-to-end optical attenuation over an optical path. Such component equipment can also be used for other measurements.
2. *Integrated test set* - When an SLS and OPM are packaged in one unit, it is called an integrated test set. Traditionally, an integrated test set is usually called an OLTS. GR-198, *Generic Requirements for Hand-Held Stabilized Light Sources, Optical Power Meters, Reflectance Meters, and Optical Loss Test Sets*, discusses OLTS equipment in depth.
3. An *Optical Time Domain Reflectometer (OTDR)* can be used to

measure optical link loss if its markers are set at the terminus points for which the fibre loss is desired. The accuracy of such a measurement can be increased if the measurement is made as a bidirectional average of the fibre. GR-196, *Generic Requirements for Optical Time Domain Reflectometer (OTDR) Type Equipment*, discusses OTDR equipment in depth.

SENSORS

The major semiconductor sensor types are Silicon (Si), Germanium (Ge) and Indium Gallium Arsenide (InGaAs). Additionally, these may be used with attenuating elements for high optical power testing, or wavelength selective elements so they only respond to particular wavelengths. These all operate in a similar type of circuit, however in addition to their basic wavelength response characteristics, each one has some other particular characteristics:

- Si detectors tend to saturate at relatively low power levels, and they are only useful in the visible and 850 nm bands.
- Ge detectors saturate at the highest power levels, but have poor low power performance, poor general linearity over the entire power range, and are generally temperature sensitive. They are only marginally accurate for "1550 nm" testing, due to a combination of temperature and wavelength affecting responsivity at *e.g.*. 1580 nm, however they provide useful performance over the commonly used 850 / 1300 / 1550 nm wavelength bands, so they are extensively deployed where lower accuracy is acceptable. Other limitations include: non-linearity at low power levels, and poor responsivity uniformity across the detector area.
- InGaAs detectors saturate at intermediate levels. They offer generally good performance, but are often very wavelength sensitive around 850 nm. So they are largely used for singlemode fibre testing at 1270 - 1650 nm.

An important part of an optical power meter sensor, is the fibre optic connector interface. Careful optical design is required to avoid significant accuracy problems when used with the wide variety of fibre types and connectors typically encountered. Another important component, is the sensor input amplifier. This needs very careful design to avoid significant performance degradation over a wide range of conditions.

POWER MEASURING RANGE

A typical OPM measures accurately under most conditions from about 0 dBm (1 milli Watt) to about -50 dBm (10 nano Watt), although the display range may be larger. Above 0 dBm is considered "high power", and specially adapted units may measure up to nearly + 30 dBm (1 Watt). Below -50 dBm is "low power", and specially adapted units may measure as low as -110 dBm. Irrespective of power meter specifications, testing below about -50 dBm tends to be sensitive to stray ambient light leaking into fibers or connectors. So when testing at "low power", some sort of test range / linearity verification (easily done with attenuators) is advisable. At low power levels, optical signal measurements tend to become noisy, so meters may become very slow due to use of a significant amount of signal averaging.

CALIBRATION AND ACCURACY

Optical Power Meter calibration and accuracy is a contentious issue. The accuracy of most primary reference standards (*e.g.*. Weight, Time, Length, Volt etc.) is known to a high accuracy, typically of the order of 1 part in a billion. However the optical power standards maintained by NIST, are only defined to about one part in a thousand. By the time this accuracy has been further degraded through successive links, instrument calibration accuracy is usually only a few per cent. The most accurate field optical power meters claim 1 per cent calibration accuracy. Comparatively, this is orders of magnitude less accurate than a typical electrical voltmeter.

Further, the in-use accuracy achieved is usually significantly lower than the claimed calibration accuracy, by the time additional factors are taken into account. In typical field applications, factors may include: ambient temperature, optical connector type, wavelength variations, linearity variations, beam geometry variations, detector saturation.Therefore, achieving a good level of practical instrument accuracy and linearity is something that requires considerable design skill, and care in manufacturing.

EXTENDED SENSITIVITY METERS

A class of laboratory power meters has an extended sensitivity, of the order of -110 dBm. This is achieved by using a very small detector and lens combination, and also a mechanical light chopper at typically 270 Hz, so the meter actually measures AC light. This eliminates unavoidable dc electrical drift effects. If the light chopping is synchronized with an appropriate synchronous (or "lock-in") amplifier, further sensitivity gains are achieved. In practice, such instruments usually achieve lower absolute accuracy due to the small detector diode, and for the same reason, may only be accurate when coupled with singlemode fibre. Occasionally such an instrument may have a cooled detector, though with the modern abandonment of Germanium sensors, and the introduction of InGaAs sensors, this is now increasingly uncommon.

PULSE POWER MEASUREMENT

Optical power meters usually display time averaged power. So for pulse measurements, the signal duty cycle must be known to calculate the peak power value. However, the instantaneous peak power must be less than the maximum meter reading, or the detector may saturate, resulting in a wrong average readings. Also, at low pulse repetition rates, some meters with data or tone detection may produce improper or no readings. A class of “high power” meters has some type of optical attenuating element in front of the detector, typically allowing about a 20 dB increase in maximum power reading. Above this level, an entirely different class of “laser power meter” instrument is used, usually based on thermal detection.

COMMON FIBRE OPTIC TEST APPLICATIONS

- Measuring the absolute power in a fibre optic signal. For this application, the power meter needs to be properly calibrated at the wavelength being tested, and set to this wavelength.
- Measuring the optical loss in a fibre, in combination with a suitable stable light source. Since this is a relative test, accurate calibration is not a particular requirement, unless two or more meters are being used due to distance issues. If a more complex two-way loss test is performed, then power meter calibration can be ignored, even when two meters are used.
- Some instruments are equipped for optical test tone detection, to assist in quick cable continuity testing. Standard test tones are usually 270 Hz, 1 kHz, 2 kHz. Some units can also determine one of 12 tones, for ribbon fibre continuity testing.

TEST AUTOMATION

Typical test automation features usually apply to loss testing applications, and include:

- The ability to set the unit to read 0 dB at a reference power level, typically the test source.
- The ability to store readings into internal memory, for subsequent recall and download to a computer.
- The ability to synchronize the wavelength with a test source, so that the meter sets to the source wavelength. This requires a specifically matched source. There simplest way of achieving this, is by recognizing a test tone, but the better way is by transfer of data. The data method has benefits that the source can send additional useful data such as nominal source power level, serial number etc.

WAVELENGTH-SELECTIVE METERS

An increasingly common special-purpose OPM, commonly called a “PON Power Meter” is designed to hook into a live PON (Passive Optical Network) circuit, and simultaneously test the optical power in different directions and wavelengths. This unit is essentially a triple power meter, with a collection of wavelength filters and optical couplers. Proper calibration is complicated by the varying duty cycle of the measured optical signals. It may have a simple pass/ fail display, to facilitate easy use by operators with little expertise.

OPTICAL TIME-DOMAIN REFLECTOMETER

An optical time-domain reflectometer (OTDR) is an optoelectronic instrument used to characterize an optical fibre. An OTDR is the optical equivalent of an electronic time domain reflectometer. It injects a series of optical pulses into the fibre under test. It also extracts, from the same end of the fibre, light that is scattered (Rayleigh backscatter) or reflected back from points along the fibre. (This is equivalent to the way that an electronic time-domain reflectometer measures reflections caused by changes in the impedance of the cable under test.) The strength of the return pulses is measured and integrated as a function of time, and plotted as a function of fibre length.

RELIABILITY AND QUALITY OF OTDR EQUIPMENT

The reliability and quality of an OTDR should be based on its accuracy, measurement range, ability to resolve and measure closely spaced events, measurement speed, and ability to perform satisfactorily under various environmental extremes and after various types of physical abuse. In addition to cost, the instrument should be rated on features provided, size, weight, and ease of use.

Accuracy is defined as the correctness of the measurement (*i.e.*, the difference between the measured value and the true value of the event being measured).

The *measurement range* of the OTDR is defined as the maximum attenuation that can be placed between the instrument and the event being measured, for which the instrument will still be able to measure the event within acceptable accuracy limits.

Instrument resolution is a measure of how close two events can be spaced and still be recognized as two separate events. The duration of the measurement pulse and the data sampling interval create a resolution limitation for OTDRs. The shorter the pulse duration and the shorter the data sampling interval, the better the instrument resolution, but the shorter the measurement range. Resolution is also often limited when powerful reflections return to the OTDR and temporarily overload the detector. When this occurs, some time is required before the instrument can resolve a second fibre event. Some OTDR manufacturers use a "masking" procedure to improve resolution. The procedure shields or "masks" the detector from high-power fibre reflections, preventing detector overload and eliminating the need for detector recovery. Industry requirements for the reliability and quality of OTDRs are in GR-196, *Generic Requirements for Optical Time Domain Reflectometer (OTDR) Type Equipment.*

TYPES OF OTDR-LIKE TEST EQUIPMENT

The common types of OTDR-like test equipment are:

- Full-feature OTDR
- Hand-held OTDR
- Fibre Break Locator
- RTU in RFTSs

The equipment is summarized below, and detailed in GR-196, Generic Requirements for Optical Time Domain Reflectometer (OTDR) Type Equipment.

- Full-feature OTDR : Full-feature OTDRs are traditional, optical time domain reflectometers. They are feature-rich and usually larger, heavier, and less portable than either the hand-held OTDR or the fibre break locator. Despite being characterized as large, their size and weight is only a fraction of that of early generation OTDRs. Often a full-feature OTDR has a main frame that can be fitted with multi-function plug-in units to perform many fibre measurement tasks. Larger colour displays are common. The full-feature OTDR often has a greater measurement range than the other types of OTDR-like equipment. Often it is used in laboratories and in the field for difficult fibre measurements. Most full-feature OTDRs are powered from AC and/or a battery.
- Hand-held OTDR and Fibre break locator : Hand-held (formerly mini) OTDRs and fibre break locators are designed to troubleshoot fibre networks in a field environment, often using battery power. The two types of instruments cover the spectrum of approaches to fibre optic plant taken by communication providers. Hand-held, inexpensive (compared to full-feature) OTDRs are intended to be easy-to-use, light-weight, sophisticated OTDRs that collect field data and perform rudimentary data analysis. They may be less feature rich than full-feature OTDRs. Often they can be used in conjunction with PC-based software to perform data collection and sophisticated data analysis. Hand-held OTDRs are commonly used to measure fibre links and locate fibre breaks, points of high loss, high reflectance, end-to-end loss, and Optical Return Loss (ORL). Fibre break locators are intended to be low-cost instruments specifically designed to determine the location of a catastrophic fibre event, *e.g.*., fibre break, point of high reflectance, or high loss. The fibre break locator is an opto-electronic tape measure designed to measure only distance to catastrophic fibre events. In general, hand-held OTDRs and fibre break locators are lighter and smaller, simpler to operate, and more likely to employ battery power than full-feature OTDRs. The intent with hand-held OTDRs and fibre break locators

is to be inexpensive enough for field technicians to be equipped with one as part of a standard tool kit.

- Remote Test Unit (RTU) : The RTU is the testing module of the RFTS described in GR-1295, *Generic Requirements for Remote Fibre Testing Systems (RFTSS)*. An RFTS enables fibre to be automatically tested from a central location. A central computer is used to control the operation of OTDR-like test components located at key points in the fibre network. The test components scan the fibre to locate problems. If a problem is found, its location is noted and the appropriate personnel are notified to begin the repair process. The RFTS can also provide direct access to a database that contains historical information of the OTDR fibre traces and any other fibre records for the physical fibre plant. Since OTDRs and OTDR-like equipment have many uses in the communications industry, operating environments vary widely, both indoors and outdoors. Most often, however, these test sets are operated in controlled environments, accessing the fibers at their termination points on fibre distribution frames. Indoor environments include controlled areas such as central offices (COs), equipment huts, or Controlled Environment Vaults (CEVs). Use in outside environments is rarer, but may include use in a manhole, aerial platform, open trench, or splicing vehicle.

OTDR DATA FORMAT

In the late 1990s, OTDR industry representatives and the OTDR user community developed a unique data format to store and analyze OTDR fibre data. This data was based on the specifications in GR-196, Generic Requirements for Optical Time Domain Reflectometer (OTDR) Type Equipment, referenced above. The goal was for the data format to be truly universal, in that it was intended to be implemented by all OTDR manufacturers. OTDR suppliers developed the software to implement the data format. As they proceeded, they identified inconsistencies in the format, along with areas of misunderstanding among users.

From 1997 to 2000, a group of OTDR supplier software specialists attempted to resolve problems and inconsistencies in what was then called the "Bellcore" OTDR Data Format. This group, called the OTDR Data Format Users Group (ODFUG), made progress. Since then, many OTDR developers continued to work with other developers to solve individual interaction problems and enable cross use between manufacturers.

In 2011, Telcordia decided to compile industry comments on this data format into one document entitled, SR-4731, *Optical Time Domain Reflectometer (OTDR) Data Format*. This Special Report (SR) summarizes the state of the Bellcore OTDR Data Format, renaming it as the Telcordia OTDR Data Format.

The data format is intended for all OTDR-related equipment designed to save trace data and analysis information. Initial implementations require standalone software to be provided by the OTDR supplier to convert existing OTDR trace files to the SR-4731 data format and to convert files from this universal format to a format that is usable by their older OTDRs. This file conversion software can be developed by the hardware supplier, the end user, or a third party. This software also provides backward compatibility of the OTDR data format with existing equipment.

The SR-4731 format describes binary data. While text information is contained in several fields, most numbers are represented as either 16-bit (2-byte) or 32-bit (4-byte) signed or unsigned integers stored as binary images. Byte ordering in this file format is explicitly low-byte ordering, as is common on Intel® processor-based machines. String fields are terminated with a zero byte "\0". OTDR waveform data are represented as short, unsigned integer data uniformly spaced in time, in units of decibels (dB) times 1000, referenced to the maximum power level. The maximum power level is set to zero, and all waveform data points are assumed to be zero or negative (the sign bit is implied), so that the minimum power level in this format is -65.535 dB, and the minimum resolution between power level steps is 0.001 dB. In some cases, this will not provide sufficient power range to represent all waveform points. For this reason, the use of a scale factor has been introduced to expand the data point power range.

OPTOMECHANICS

Optomechanics can refer to:

- The sub-field of physics involving the study of the interaction of electromagnetic radiation (photons) with mechanical systems via radiation pressure or
- The manufacture and maintenance of optical parts and devices.

This second definition includes:

- Production of optomechanical parts, such as
  - Mirror mounts
  - Optical mounts
  - Translation stages
  - Rotary and kinematic stages
  - Fibre aligners
  - Rails
  - Pedestals and posts
  - Micrometers, screws and screw sets
- Design and packaging of compact and rugged optical trains
- Optomechanical design and integration with exterior package
- Manufacture and maintenance of fibre optic materials

OVALITY

In telecommunications and fibre optics, ovality or non-circularity is the degree of deviation from perfect circularity of the cross section of the core or cladding of the fibre.

The cross-sections of the core and cladding are assumed to a first approximation to be elliptical. Quantitatively, the ovality of either the core or cladding is expressed as $\frac{2(a-b)}{a+b}$, where a is the length of the major axis and b is the length of the minor axis. The dimensionless quantity so obtained may be multiplied by 100 to express ovality as a percentage. Alternatively, ovality of the core or cladding may be specified by a tolerance field consisting of two concentric circles, within which the cross section boundaries must lie. In measurements, ovality is the amount of out-of-roundness of a hole or cylindrical part in the typical form of an oval.

POWER-LAW INDEX PROFILE

For optical fibers, a power-law index profile is an index of refraction profile characterized by

$$n(r) = \begin{cases} n_1\sqrt{1-2\Delta\left(\frac{r}{\alpha}\right)^g} & r \le \alpha \\ n_1\sqrt{1-2\Delta} & r \ge \alpha \end{cases}$$

where $\Delta = \frac{n_1^2 - n_2^2}{2n_1^2}$,

and $n(r)$ is the nominal refractive index as a function of distance from the fibre axis, n_1 is the nominal refractive index on axis, n_2 is the refractive index of the cladding, which is taken to be homogeneous ($n(r) = n_2$ for $r \ge \alpha$), α is the core radius, and g is a parameter that defines the shape of the profile. α is often used in place of g. Hence, this is sometimes called an alpha profile.

For this class of profiles, multimode distortion is smallest when g takes a particular value depending on the material used. For most materials, this optimum value is approximately 2. In the limit of infinite g, the profile becomes a step-index profile.

RADIATION ANGLE

In fibre optics, the radiation angle is half the vertex angle of the cone of light emitted at the exit face of an optical fibre. The cone boundary is usually defined (a) by the angle at which the far-field irradiance has decreased to a specified fraction of its maximum value or (b) as the cone within which there is a specified fraction of the total radiated power at any point in the far field.

RADIATION MODE

For an optical fibre or waveguide, a radiation mode or unbound mode is a mode which is not confined by the fibre core. Such a mode has fields that are transversely oscillatory everywhere external to the waveguide, and exists even at the limit of zero wavelength.

Specifically, a radiation mode is one for which

$$\beta = \sqrt{n^2(a)k^2 - (l/a)^2}$$

where β is the imaginary part of the axial propagation constant, integer l is the azimuthal index of the mode, $n(r)$ is the refractive index at radius r, a is the core radius, and k is the free-space wave number, $k = 2\pi/\lambda$, where λ is the wavelength. Radiation modes correspond to refracted rays in the terminology of geometric optics.

RADIATION PATTERN

In the field of antenna design the term radiation pattern (or antenna pattern or far-field pattern) refers to the *directional* (angular) dependence of the strength of the radio waves from the antenna or other source.

Particularly in the fields of fibre optics, lasers, and integrated optics, the term radiation pattern may also be used as a synonym for the near-field pattern or Fresnel pattern. This refers to the *positional* dependence of the electromagnetic field in the near-field, or Fresnel region of the source. The near-field pattern is most commonly defined over a plane placed in front of the source, or over a cylindrical or spherical surface enclosing it.

The far-field pattern of an antenna may be determined experimentally at an antenna range, or alternatively, the near-field pattern may be found using a near-field scanner, and the radiation pattern deduced from it by computation. The far-field radiation pattern can also be calculated from the antenna shape by computer programs such as NEC. Other software, like HFSS can also compute the near field.

The far field radiation pattern may be represented graphically as a plot of one of a number of related variables, including; the field strength at a constant (large) radius (an amplitude pattern or field pattern), the power per unit solid angle (power pattern) and the directive gain. Very often, only the relative amplitude is plotted, normalized either to the amplitude on the antenna boresight, or to the total radiated power. The plotted quantity may be shown on a linear scale, or in dB. The plot is typically represented as a three-dimensional graph (as at right), or as separate graphs in the vertical plane and horizontal plane. This is often known as a polar diagram.

RECIPROCITY

It is a fundamental property of antennas that the receiving pattern (sensitivity as a function of direction) of an antenna when used for receiving is identical to the far-field radiation pattern of the antenna when used for transmitting. This is a consequence of the reciprocity theorem of electromagnetics and is proved below. Therefore in discussions of radiation patterns the antenna can be viewed as either transmitting or receiving, whichever is more convenient.

TYPICAL PATTERNS

Since electromagnetic radiation is dipole radiation, it is not possible to build an antenna that radiates equally in all directions, although such a hypothetical isotropic antenna is used as a reference to calculate antenna gain. The simplest antennas, monopole and dipole antennas, consist of one or two straight metal rods along a common axis.

These axially symmetric antennas have radiation patterns with a similar symmetry, called omnidirectional patterns; they radiate equal power in all directions perpendicular to the antenna, with the power varying only with the angle to the axis, dropping off to zero on the antenna's axis. This illustrates the general principle that if the shape of an antenna is symmetrical, its radiation pattern will have the same symmetry.

In most antennas, the radiation from the different parts of the antenna interferes at some angles. This results in zero radiation at certain angles where the radio waves from the different parts arrive out of phase, and local maxima of radiation at other angles where the radio waves arrive in phase. Therefore the radiation plot of most antennas shows a pattern of maxima called "lobes" at various angles, separated by "*nulls*" at which the radiation goes to zero.

The larger the antenna is compared to a wavelength, the more lobes there will be. In a directive antenna in which the objective is to direct the radio waves in one particular direction, the lobe in that direction is larger than the others; this is called the "*main lobe*". The axis of maximum radiation, passing through the center of the main lobe, is called the "*beam axis*" or *boresight axis*". In some antennas, such as split-beam antennas, there may exist more than one major lobe. A minor lobe is any lobe except a major lobe.

The other lobes, representing unwanted radiation in other directions, are called "*side lobes*". The side lobe in the opposite direction (180°) from the main lobe is called the "*back lobe*". Usually it refers to a minor lobe that occupies the hemisphere in a direction opposite to that of the major (main) lobe.

Minor lobes usually represent radiation in undesired directions, and they should be minimized. Side lobes are normally the largest of the minor lobes. The level of minor lobes is usually expressed as a ratio of the power density in the lobe in question to that of the major lobe. This ratio is often termed the side lobe ratio or side lobe level. Side lobe levels of "20 dB or smaller are usually not desirable in many applications. Attainment of a side lobe level smaller than

"30 dB usually requires very careful design and construction. In most radar systems, for example, low side lobe ratios are very important to minimize false target indications through the side lobes.

PROOF OF RECIPROCITY

Here, we present a common simple proof limited to the approximation of two antennas separated by a large distance compared to the size of the antenna, in a homogeneous medium. The first antenna is the test antenna whose patterns are to be investigated; this antenna is free to point in any direction. The second antenna is a reference antenna, which points rigidly at the first antenna.

Each antenna is alternately connected to a transmitter having a particular source impedance, and a receiver having the same input impedance (the impedance may differ between the two antennas).

It will be assumed that the two antennas are sufficiently far apart that the properties of the transmitting antenna are not affected by the load placed upon it by the receiving antenna. Consequently, the amount of power transferred from the transmitter to the receiver can be expressed as the product of two independent factors; one depending on the directional properties of the transmitting antenna, and the other depending on the directional properties of the receiving antenna. For the transmitting antenna, by the definition of gain, G, the radiation power density at a distance r from the antenna (*i.e.* the power passing through unit area) is

$$W(\theta,\Phi) = \frac{G(\theta,\Phi)}{4\pi r^2} Pt$$

Here, the arguments θ and Φ indicate a dependence on direction from the antenna, and Pt stands for the power the transmitter would deliver into a matched load. The gain G may be broken down into three factors; the antenna gain (the directional redistribution of the power), the radiation efficiency (accounting for ohmic losses in the antenna), and lastly the loss due to mismatch between the antenna and transmitter. Strictly, to include the mismatch, it should be called the realized gain, but this is not common usage.

For the receiving antenna, the power delivered to the receiver is

$$P_r = A(\theta,\Phi)W \, .$$

Here W is the power density of the incident radiation, and A is the antenna aperture or effective area of the antenna (the area the antenna would need to occupy in order to intercept the observed captured power). The directional arguments are now relative to the receiving antenna, and again A is taken to include ohmic and mismatch losses.

Putting these expressions together, the power transferred from transmitter to receiver is

$$P_r = A\frac{G}{4\pi r^2}Pt\,,$$

where G and A are directionally dependent properties of the transmitting and receiving antennas respectively. For transmission from the reference antenna (2), to the test antenna (1), that is

$$P_{1r} = A_1(\theta,\Phi)\frac{G_2}{4\pi r^2}P_{2t}\,,$$

and for transmission in the opposite direction

$$P_{2r} = A_2\frac{G_1(\theta,\Phi)}{4\pi r^2}P_{1t}\,.$$

Here, the gainG_2 and effective area A_2 of antenna 2 are fixed, because the orientation of this antenna is fixed with respect to the first.

Now for a given disposition of the antennas, the reciprocity theorem requires that the power transfer is equally effective in each direction, *i.e.*

$$\frac{P_{1r}}{P_{2t}} = \frac{P_{2r}}{P1t}\,,$$

whence

$$\frac{A_1(\theta,\Phi)}{G_1(\theta,\Phi)} = \frac{A_2}{G_2}\,.$$

But the right hand side of this equation is fixed (because the orientation of antenna 2 is fixed), and so

$$\frac{A_1(\theta,\Phi)}{G_1(\theta,\Phi)} = \text{constant}\,,$$

i.e. the directional dependence of the (receiving) effective aperture and the (transmitting) gain are identical (QED). Furthermore, the constant of proportionality is the same irrespective of the nature of the antenna, and so must be the same for all antennas. Analysis of a particular antenna (such as a Hertzian dipole), shows that this constant is $\frac{\lambda^2}{4\pi}$, where λ is the free-space wavelength. Hence, for any antenna the gain and the effective aperture are related by

$$A(\theta,\Phi) = \frac{\lambda^2 G(\theta,\Phi)}{4\pi}\,.$$

Even for a receiving antenna, it is more usual to state the gain than to specify the effective aperture. The power delivered to the receiver is therefore more usually written as

$$P_r = \frac{\lambda^2 G_r G_t}{(4\pi r)^2} P_t$$

The effective aperture is however of interest for comparison with the actual physical size of the antenna.

Practical consequences

- When determining the pattern of a receiving antenna by computer simulation, it is not necessary to perform a calculation for every possible angle of incidence. Instead, the radiation pattern of the antenna is determined by a single simulation, and the receiving pattern inferred by reciprocity.
- When determining the pattern of an antenna by measurement, the antenna may be either receiving or transmitting, whichever is more convenient.

RAY (OPTICS)

In optics a ray is an idealized model of light, obtained by choosing a line that is perpendicular to the wavefronts of the actual light, and that points in the direction of energy flow. Rays are used to model the propagation of light through an optical system, by dividing the real light field up into discrete rays that can be computationally propagated through the system by the techniques of ray tracing.

This allows even very complex optical systems to be analyzed mathematically or simulated by computer. Ray tracing uses approximate solutions to Maxwell's equations that are valid as long as the light waves propagate through and around objects whose dimensions are much greater than the light's wavelength. Ray theory does not describe phenomena such as interference and diffraction, which require wave theory (involving the relative phase of the rays).

A light ray is a line or curve that is perpendicular to the light's wavefronts (and is therefore collinear with the wave vector). Light rays bend at the interface between two dissimilar media and may be curved in a medium in which the refractive index changes. Geometric optics describes how rays propagate through an optical system. Objects to be imaged are treated as collections of independent point sources, each producing spherical wavefronts and corresponding outward rays. Rays from each object point can be mathematically propagated to locate the corresponding point on the image. A slightly more rigorous definition of a light ray follows from Fermat's principle, which states that the path taken between two points by a ray of light is the path that can be traversed in the least time.

SPECIAL RAYS

There are many special rays that are used in optical modelling to analyze an optical system. These are defined and described below, grouped by the type of system they are used to model.

Interaction with surfaces

- An incident ray is a ray of light that strikes a surface. The angle between this ray and the perpendicular or normal to the surface is the angle of incidence.
- The reflected ray corresponding to a given incident ray, is the ray that represents the light reflected by the surface. The angle between the surface normal and the reflected ray is known as the angle of reflection. The Law of Reflection says that for a specular (non-scattering) surface, the angle of reflection always equals the angle of incidence.
- The refracted ray or transmitted ray corresponding to a given incident ray represents the light that is transmitted through the surface. The angle between this ray and the normal is known as the angle of refraction, and it is given by Snell's Law. Conservation of energy requires that the power in the incident ray must equal the sum of the power in the refracted ray, the power in the reflected ray, and any power absorbed at the surface.
- If the material is birefringent, the refracted ray may split into ordinary and extraordinary rays, which experience different indexes of refraction when passing through the birefringent material.

Optical systems

- A meridional ray or tangential ray is a ray that is confined to the plane containing the system's optical axis and the object point from which the ray originated.
- A skew ray is a ray that does not propagate in a plane that contains both the object point and the optical axis. Such rays do not cross the optical axis anywhere, and are not parallel to it.
- The marginal ray (sometimes known as an *a ray* or a *marginal axial ray*) in an optical system is the meridional ray that starts at the point where the object crosses the optical axis, and touches the edge of the aperture stop of the system. This ray is useful, because it crosses the optical axis again at the locations where an image will be formed. The distance of the marginal ray from the optical axis at the locations

of the entrance pupil and exit pupil defines the sizes of each pupil (since the pupils are images of the aperture stop).

- The principal ray or chief ray (sometimes known as the *b ray*) in an optical system is the meridional ray that starts at the edge of the object, and passes through the center of the aperture stop. This ray crosses the optical axis at the locations of the pupils. As such chief rays are equivalent to the rays in a pinhole camera. The distance between the chief ray and the optical axis at an image location defines the size of the image. The marginal and chief rays together define the Lagrange invariant, which characterizes the throughput or etendue of the optical system. Some authors define a "principal ray" for *each* object point. The principal ray starting at a point on the edge of the object may then be called the *marginal principal ray*.
- A sagittal ray or transverse ray from an off-axis object point is a ray that propagates in the plane that is perpendicular to the meridional plane and contains the principal ray. Sagittal rays intersect the pupil along a line that is perpendicular to the meridional plane for the ray's object point and passes through the optical axis. If the axis direction is defined to be the z axis, and the meridional plane is the y-z plane, sagittal rays intersect the pupil at $y_p=0$. The principal ray is both sagittal and meridional. All other sagittal rays are skew rays.
- A paraxial ray is a ray that makes a small angle to the optical axis of the system, and lies close to the axis throughout the system. Such rays can be modeled reasonably well by using the paraxial approximation. When discussing ray tracing this definition is often reversed: a "paraxial ray" is then a ray that is modeled using the paraxial approximation, not necessarily a ray that remains close to the axis.
- A finite ray or real ray is a ray that is traced without making the paraxial approximation.
- A parabasal ray is a ray that propagates close to some defined "base ray" rather than the optical axis. This is more appropriate than the paraxial model in systems that lack symmetry about the optical axis. In computer modeling, parabasal rays are "real rays", that is rays that are treated without making the paraxial approximation. Parabasal rays about the optical axis are sometimes used to calculate first-order properties of optical systems.

Fibre optics

- A meridional ray is a ray that passes through the axis of an optical fibre.
- A skew ray is a ray that travels in a non-planar zig-zag path and never crosses the axis of an optical fibre.

- A guided ray, bound ray, or trapped ray is a ray in a multi-mode optical fibre, which is confined by the core. For step index fibre, light entering the fibre will be guided if it makes an angle with the fibre axis that is less than the fiber's acceptance angle.
- A leaky ray or tunneling ray is a ray in an optical fibre that geometric optics predicts would totally reflect at the boundary between the core and the cladding, but which suffers loss due to the curved core boundary.

RECOATING

Recoating is the process of restoring the primary coating to stripped optical fibre sections after fusion splicing. In the recoating process, the spliced fibre is restored to its original shape and strength, using a recoater. The stripped fibre section is recoated by filling a recoating resin, usually acrylate into transparent moulds. The resin is then cured with UV light. It is often desirable to perform a proof-test after recoating, to ensure that the splice is strong enough to survive handling, packaging and extended use.

Recoated splices can usually be coiled into a tight radius because the recoated fibre section is just as flexible as the original polymer coating. Recoats are *e.g.*. employed in the assembly of undersea optical fibre cables where high fusion splice strength is already necessitated by stringent reliability requirements. Another reason optical fibre recoating is attractive in submarine communications cabling is that the cross section of recoated fibers matches that of the original fibre. The first commercial optical fibre recoaters were developed by the Swedish firms Nyfors Teknologi AB and LM Ericsson in the late 1980s for the telecommunications industry.

REFERENCE SURFACE

In fibre optic technology, a reference surface is that surface of an optical fibre that is used to contact the transverse-alignment elements of a component such as a connector or mechanical splice. For telecommunications-grade fibers, the reference surface is the outer surface of the cladding. For plastic-clad silica (PCS) fibers, which have a strippable polymer cladding (not to be confused with the polymer overcoat of an all-silica fibre), the reference surface may be the core.

REFRACTIVE INDEX CONTRAST

Refractive index contrast, in an optical fibre, is a measure of the relative difference in refractive index of the core and cladding. Refractive index contrast, Ä, is given by $Ä = (n_1 - n_2)/(2n_1)$, where n_1 is the maximum refractive index in the core and n_2 is the refractive index of the homogeneous cladding. Normal optical fibers have very low refractive index contrast(Ä<<1)hence are weakly guided medium. The weak guiding will cause more of the Electrical field to "leak" and travel through the cladding(as evanescent waves) as compared to the strongly guided waveguides.

SOLAR PHOTONICS

Solar photonics is a science on the boundary between photonics/optical communication and solar energy. Basically the idea is to distribute solar energy though optical fibres or other photonic guides (liquids). The term is first found in the work by a research team at the Department of Solar Energy and Environmental Physics at the Ben-Gurion University of the Negev, Israel:

- Solar photonics refers to the exploitation of the inherent photonic or quantum-mechanical value of sunlight... there are instances for which immense photon densities are required, as well as the wavelengths present in the solar spectrum.—Gordon Feuermann, *Online pamphlet "Department of solar energy and environmental physics" page 13 (January 2000)*

Examples mentioned in the referred publication are solar surgery, concentration of photonic power generation and solar photonics for nanotechnology.

Research topics:

- The wide spectrum of solar light
- The high power of concentrated solar light
- Solar light concentration
- Solar laser generation
- Artificial photo-synthesis
- Photo-voltaic or nanoantenna form of power generation
- Diffusion for lighting
- Diffused and beamed form of collection
- Hybride and simultaneous applications

Optical fibres are not designed to the given conditions. Research will have to find solutions to enable high power light transmission through optical fibres and other devices (*e.g.*. liquids).

Applications:

- Potential applications of solar photonics are indoor natural lighting using optical fibers or light pipes, high efficient improved solar power generation, clean fuel such as hydrogen generation and solar chemistry.

SPLIT-STEP METHOD

In numerical analysis, the split-step (Fourier) method is a pseudo-spectral numerical method used to solve non-linear partial differential equations like the non-linear Schrödinger equation. The name arises for two reasons. First, the method relies on computing the solution in small steps, and treating the linear and the non-linear steps separately (see below). Second, it is necessary to Fourier transform back and forth because the linear step is made in the frequency domain while the non-linear step is made in the time domain.

An example of usage of this method is in the field of light pulse propagation in optical fibers, where the interaction of linear and non-linear mechanisms makes it difficult to find general analytical solutions. However, the split-step method provides a numerical solution to the problem.

DESCRIPTION OF THE METHOD

Consider, for example, the non-linear Schrödinger equation

$$\frac{\partial A}{\partial z} = -\frac{i\beta_2}{2}\frac{\partial^2 A}{\partial t^2} + i\gamma |A|^2 A = [\hat{D} + \hat{N}]A,$$

where $A(t, z)$ describes the pulse envelope in time t at the spatial position z. The equation can be split into a linear part,

$$\frac{\partial A_D}{\partial z} = -\frac{i\beta_2}{2}\frac{\partial^2 A}{\partial t^2} = \hat{D}A,$$

and a non-linear part,

$$\frac{\partial A_N}{\partial z} = i\gamma |A|^2 A = \hat{N}A,$$

Both the linear and the non-linear parts have analytical solutions, but the non-linear Schrödinger equation containing both parts does not have a general analytical solution.

However, if only a 'small' step h is taken along z, then the two parts can be treated separately with only a 'small' numerical error. One can therefore first take a small non-linear step,

$$A_N(t, z+h) = \exp[i\gamma |A|^2 h]\, A(t, z),$$

using the analytical solution.

The dispersion step has an analytical solution in the frequency domain, so it is first necessary to Fourier transform A_N using

$$\tilde{A}_N(\omega, z) = \int_{-\infty}^{\infty} A_N(t, z) \exp[i(\omega - \omega_0)t]dt,$$

where ω_0 is the center frequency of the pulse. It can be shown that using the above definition of the Fourier transform, the analytical solution to the linear step, commuted with the frequency domain solution for the non-linear step, is

$$\tilde{A}(\omega, z+h) = \exp\left[\frac{i\beta_2}{2}(\omega - \omega_0)^2 h\right]\tilde{A}_N(\omega, z)$$

By taking the inverse Fourier transform of $\tilde{A}(\omega, z+h)$ one obtains $\tilde{A}(t, z+h)$; the pulse has thus been propagated a small step h. By repeating the above N times, the pulse can be propagated over a length of Nh.

The above shows how to use the method to propagate a solution forward in space; however, many physics applications, such as studying the evolution of a wave packet describing a particle, require one to propagate the solution forward in time rather than in space. The non-linear Schrödinger equation, when used to govern the time evolution of a wave function, takes the form

$$i\hbar\frac{\partial\psi}{\partial t} = -\frac{\hbar^2}{2m}\frac{\partial^2\psi}{\partial x^2} + \gamma|\psi|^2\psi = [\hat{D} + \hat{N}]\psi,$$

where $\psi(x, t)$ describes the wave function at position x and time t. Note that $\hat{D} = -\frac{\hbar^2}{2m}\frac{\partial^2}{\partial x^2}$ and $\hat{N} = \gamma|\psi|^2$, and that m is the mass of the particle and $\hbar$ is Planck's constant over 2π.

The formal solution to this equation is a complex exponential, so we have that

$$\psi(x,t) = e^{it(\hat{D}+\hat{N})}\psi(x,0).$$

Since $\hat{D}$ and $\hat{N}$ are operators, they do not in general commute. However, the Baker-Hausdorff formula can be applied to show that the error from treating them as if they do will be of order dt^2 if we are taking a small but finite time step dt. We therefore can write

$$\psi(x,t+dt) \approx e^{idt\hat{D}}e^{idt\hat{N}}\psi(x,t).$$

The part of this equation involving $\hat{N}$ can be computed directly using the wave function at time t, but to compute the exponential involving $\hat{D}$ we use the fact that in frequency space, the partial derivative operator can be converted into a number by substituting ik for $\frac{\partial}{\partial x}$, where k is the frequency (or more properly, wave number, as we are dealing with a spatial variable and thus transforming to a space of spatial frequencies—*i.e.* wave numbers) associated with the Fourier transform of whatever is being operated on. Thus, we take the Fourier transform of

$$e^{idt\hat{N}}\psi(x,t),$$

recover the associated wave number, compute the quantity

$$e^{-idtk^2},$$

and use it to find the product of the complex exponentials involving $\hat{N}$ and $\hat{D}$ in frequency space as below:

$$e^{-idtk^2} F[e^{idt\hat{N}} \psi(x,t)],$$

where F denotes a Fourier transform. We then inverse Fourier transform this expression to find the final result in physical space, yielding the final expression

$$\psi(x,t+dt) = F^{-1}[e^{-idtk^2} F[e^{idt\hat{N}} \psi(x,t)]].$$

A variation on this method is the symmetrized split-step Fourier method, which takes half a time step using one operator, then takes a full-time step with only the other, and then takes a second half time step again with only the first. This method is an improvement upon the generic split-step Fourier method because its error is of order dt^3 for a time step dt. The Fourier transforms of this algorithm can be computed relatively fast using the *fast Fourier transform (FFT)*. The split-step Fourier method can therefore be much faster than typical finite difference methods.

STRIPPING (FIBRE)

Stripping is the act of removing the protective polymer coating around optical fibre in preparation for fusion splicing. The splicing process begins by preparing both fibre ends for fusion, which requires that all protective coating is removed or stripped from the ends of each fibre. Fibre optical stripping can be done using a special stripping and preparation unit that uses hot sulphuric acid or a controlled flow of hot air to remove the coating. There are also mechanical tools used for stripping fibre which are similar to copper wire strippers. Fibre optical stripping and preparation equipment used in fusion splicing is commercially available through a small number of specialized companies, which usually also design machines used for fibre optical recoating.

SUBSTITUTION METHOD

In optical fibre technology, the substitution method is a method of measuring the transmission loss of a fibre. It consists of:

1. Using a stable optical source, at the wavelength of interest, to drive a mode scrambler, the output of which overfills (drives) a 1 to 2 meter long reference fibre having physical and optical characteristics matching those of the fibre under test,
2. Measuring the power level at the output of the reference fibre,
3. Repeating the procedure, substituting the fibre under test for the reference fibre, and
4. Subtracting the power level obtained at the output of the fibre under test from the power level obtained at the output of the reference fibre, to get the transmission loss of the fibre under test.

The substitution method has certain shortcomings with regard to its accuracy, but its simplicity makes it a popular field test method. It is conservative, in that if it were used to measure the individual losses of several long fibers, and the long fibers were concatenated, the total loss obtained (excluding splice losses) would be expected to be lower than the sum of the individual fibre losses. Some modern optical power meters have the capability to set to zero the reference level measured at the output of the reference fibre, so that the transmission loss of the fibre under test may be read out directly.

ZERO-DISPERSION WAVELENGTH

In a single-mode optical fibre, the zero-dispersion wavelength is the wavelength or wavelengths at which material dispersion and waveguide dispersion cancel one another. In all silica-based optical fibers, minimum material dispersion occurs naturally at a wavelength of approximately 1300nm. Single-mode fibers may be made of silica-based glasses containing dopants that shift the material-dispersion wavelength, and thus, the zero-dispersion wavelength, towards the minimum-loss window at approximately 1550nm. The engineering tradeoff is a slight increase in the minimum attenuation coefficient. Such fibre is called dispersion-shifted fibre.

Another way to alter the dispersion is changing the core size and the refractive indices of the material of core and cladding. Because fibre optic materials are already highly optimized for low scattering and high transparency alternative ways to change the refractive index were investigated. As a straight forward solution tapered fibers and holey fibers or photonic crystal fibers (PCF) were produced. Essentially they replace the cladding by air. This improves the contrast of refractive indices by a factor of 10. Therefore the effective index is changed, especially for longer wavelengths.

This type of refractive index change versus wavelength due to different geometry is called waveguide dispersion. As these narrow waveguides (~1-3 μm core diameter) are combined with ultrashort pulses at the zero-dispersion wavelength pulses are not instantly destroyed by dispersion. After reaching a certain peak power within the pulse the non-linear refractive index starts to play an important role leading to frequency generation processes like self-phase modulation (SPM), modulational instability, soliton generation and soliton fission, cross phase modulation (XPM) and others. All these processes generate new frequency components, meaning that input light with narrow bandwidth expands into a wide range of new colours, through a process called supercontinuum generation.

The term is also used, more loosely, in multi-mode optical fibre. There, it refers to the wavelength at which the material dispersion is minimum, *i.e. essentially* zero. This is more accurately called the minimum-dispersion wavelength.

ZERO-DISPERSION SLOPE

The rate of change of dispersion with respect to wavelength at the zero-dispersion point is called the zero-dispersion slope. Doubly and quadruply clad single-mode fibers have two zero-dispersion points, and thus two zero-dispersion slopes.

3

Textile and Apparel: Marketing Trends

This is a marketing research on the Textiles industry and can include information on the background, market structure, definitions, competitors, trends and developments of textiles and is related to other apparel, clothing, yarn, thread and fibre.

MARKET TRENDS

Global market trends in the textile industry increasingly are being determined by the transition to a quota-free textile trade environment. The quota system that has been in place since the early 1960s ended on December 31, 2004. This situation is driving a realignment of markets and a restructuring of the industry that are well under way.

The response of textile manufacturers to intensifying global competition has been threefold. First, industry product mixes are shifting among countries. There is ongoing rationalization of production to emphasize products in which the manufacturer has the greatest competitive advantages and de-emphasize products that are more vulnerable to foreign competition.

In the United States and other developed countries, this implies a de-emphasis on textile products consumed in the apparel industry and a greater emphasis on other products, such as home furnishings and industrial textiles. Second, companies continue to invest in new, more productive plant and equipment. Textile production is no longer a low-technology, labour-intensive industry enterprise.

Textile firms around the world are under constant pressure to become more efficient through technology upgrades. Third and perhaps most important, manufacturing is expanding into countries close to home markets. This is being undertaken primarily to support production sharing arrangements for apparel, sometimes in partnership with customers, suppliers, or other textile manufacturers. Textiles refer to yarns, threads and wools that can be spun, woven, tufted, tied and otherwise used to manufacture cloth. Production of textiles has been altered almost beyond recognition by mass-production and the introduction of modern manufacturing techniques.

MARKET STRUCTURE

Economies of scale in textile manufacturing are significant and create entry barriers in the market. The textile industry is more capital intensive than the clothing industry and is highly automated. It consists of spinning, weaving and finishing functions in integrated plants. For example, the approximate cost of a new fibre plant is $100 million depending on various factors. Costs of raw materials are volatile and can account for 50-60 per cent of the cost of the finished products. For purposes of hedging against supply problems, many manufacturers are backward integrated into chemical intermediaries. Forward integration into apparel and product manufacturing is also seen in industry.

In 1995, the World Trade Organization (WTO), in the Agreement on Textiles and Clothing (ATC), agreed that all quotas on textiles and clothing would disappear between WTO member countries on January 1, 2005. The expiration of ATC marked the end of quotas, limiting textile and clothing trade between the WTO members. Developing countries such as China, India and Pakistan had been the most restricted by the quotas. While India and China are likely to emerge as winners, the main losers after quota will be quota-restricted countries that had enjoyed the benefits and protection for more than 40 years. The post quota market has changed with producers already affected by significant changes in retailing. Large retailers are now buying up independent brands to give consumers more value and enhance their shopping experience. Manufacturers in developed countries are more likely to adapt by relocating operations to production centers in low wage countries.

Some of the major textile industries can be divided as follows:

- Awnings, textile
- Blankets
- Bags or sacks, textile
- Blinds, textile
- Canvas goods
- Cordage
- Elasticized fabrics
- Fabrics, textile
- Felt (except floor coverings)
- Glass fibre fabrics
- Household linen
- Lace
- Narrow fabrics
- Netting
- Rope (except wire rope)
- Sailcloth
- Sewing thread
- Soft furnishings

- String
- Tarpaulins
- Tents
- Thread
- Towels
- Trimmings, textile
- Yarns

For U.S. textile producers, the most important regional expansion and partnering are taking place with Mexico. Broad market forces, including Asian pricing pressures, have accelerated the flow of textile capital into Mexico.

INDUSTRY DEFINITIONS

- *Acrylic*: Manufactured fibre derived from polyacrylonitrile. Its major properties include a soft, wool-like hand, machine washable and dryable, excellent colour retention. Solution-dyed versions have excellent resistance to sunlight and chlorine degradation.
- *Anti-dumping duty*: An extra duty imposed on an imported product by an importing country (or group of countries, as in the case of the EU) to compensate for the dumping of goods by a foreign supplier.
- *Aramid*: The generic name for a special group of synthetic fibres (aromatic polyamide) having high strength; examples are "Kevlar" and "Twaron".
- *Calico*: Tightly woven cotton type fabric with an all-over print, usually a small floral pattern on a contrasting background colour. Common end-uses include dresses, aprons, and quilts.
- *Countervailing duty*: An extra duty imposed on an imported product by an importing country (or group of countries, as in the case of the EU) to compensate for subsidies deemed to be illegal which are given to the manufacturer of the product in the exporting country.
- *Delocalization*: The geographical move of a production unit to a low cost country. (Note that the term is increasingly being used to describe all forms of shifts in production, including foreign sourcing and subcontracting.)
- *Industrial textiles*: A category of technical textiles used as part of an industrial process, or incorporated into final products.
- *Lycra*: DuPont's brand name for its elastane or spandex fibre.
- *PET*: Polyethylene terephthalate, the most common form of polyester.
- *Polyester*: A manufactured fibre introduced in the early 1950s, and is second only to cotton in worldwide use. Polyester has high strength (although somewhat lower than nylon), excellent resiliency, and high abrasion resistance. Low absorbency allows the fibre to dry quickly.
- *Quota*: A quantitative restraint imposed by an importing country on an exporting country - or established by agreement between the

trading partners - which is designed to limit shipments of a product from the exporting to the importing country.

- *Spandex fibre*: The generic name used in the USA to denote elastane fibre. A manufactured elastomeric fibre that can be repeatedly stretched over 500 per cent without breaking, and will still recover to its original length.
- *Technical textiles*: Textile materials and products manufactured primarily for their technical performance and functional properties rather than their aesthetic or decorative characteristics. End uses include aerospace, industrial, marine, medical, military, safety and transport textiles, and geotextiles.
- *Unit production systems*: An advanced apparel manufacturing system in which a single garment is progressed through a sequence of operations. Using a unit production system, a garment is automatically transported via a computer-controlled overhead hanging system, which has been ergonomically designed to reduce the amount of handling of the garment (see also progressive bundle system).
- *Yarn*: A continuous strand of textile fibres created when a cluster of individual fibres are twisted together. These long yarns are used to create fabrics, either by knitting or weaving.
- *Yarn Numbering System*: Systems for sizing yarn fall into two basic types. The yarn number is based on the length of yarn needed to make up a specified weight. The finer the yarn, the higher the number. Cotton, wool and linen are numbered with such systems. The yarn number is based on the mass of a specified length of yarn. The finer the yarn, the lower the number. Silk, synthetic fibres and jute are numbered with such systems.
- *Textile HS codes*: Harmonized System Codes (HS Codes) are a government, regulated system used to classify products and their corresponding tariffs. The Harmonized System is an internationally developed and implemented commodity-description and coding system, on which the tariffs of most countries of the world (including the NAFTA countries) are based.
- *Manufactured Fibres*: It is important to understand that all manufactured fibres are not alike. Each fibre has a unique composition and physical properties. The U. S. Federal Trade Commission has established generic names and definitions for manufactured fibres, including acetate, acrylic, lyocell, modacrylic, nylon, polyester, polypropylene (olefin), rayon, and spandex.

SUSTAINABILITY

Sustain means "to maintain" or " to uphold" and with regard to industrial processes sustainability means establishing hose principles and practices which can help to maintain the equilibrium of nature in other words to avoid causing irreversible damage to the earth's natural resources. Moving to a greater degree of sustainability in our industrial processes and systems requires that we achieve a better balance between the social, economic and environmental aspects of textile production.

- A sustainable product is one that is manufactured in a way that respects the social elements of fair trade and human rights of the people involved in the whole of the manufacturing chain.
- A sustainable product is one that is manufactured in such a way that it has the lowest possible adverse effect on the environment *e.g.* by making the most efficient use of resources such as water and energy, and which goes the extra mile to recover raw materials, *e.g.* by the recycling of as much water as possible or by recovering the heat from wastewater discharges.
- But equally important, a sustainable product is one which can compete effectively in the global marketplace against less sustainable products. *i.e.* which offers value benefits to the consumer, and where the economic returns from its success are fairly distributed back along the supply chain.

In order to achieve the above agenda, it is important to work at each and every aspect of textile production where natural resources, energy and chemicals are consumed and emissions to air, water and land arise.

SUSTAINABLE TEXTILE PROCESSING

As we have already noted the textile industry is one of the most polluting industry sectors. A vast range and quantity of chemicals is used at every stage and the after- effects in terms of wastewater treatment and air pollution are critical to manage. There is a strong need to establish more sustainable textile processing measures in the industry.

The main objective of these measures should be to minimize and eliminate the most harmful inputs and the most polluting outputs and to reduce the level of chemical residues left on the textile. But it would be wrong to focus solely on the chemical inputs and ignore the consumption of energy and water which are the primary impacts of the textile processing industry.

In summary a sustainable approach covers the following points:

- Minimum use of resources (water and energy)
- Minimum chemical consumption
- Minimum pollution load
- Toxic chemicals eliminated from supply chain
- Harmful chemical residues eliminated from final textile

But in order to minimize the usage, it is important to measure the inputs and in order to eliminate the most harmful chemicals, it is important to know and understand what is being used. Uncontrolled or unknown inputs lead to unmanaged use of resources and uncontrolled outputs. The measurement of the parameters shown below is essential to establish an understanding of the input-output balance of the textile processing operation.

The measurement and control of these inputs and outputs can lead to:

- Improved resource productivity
- Improved eco-efficiency
- Improved cost efficiency
- Improved customer satisfaction
- Improved brand reputation

Once the understanding about the basics of processes and chemicals involved with the supply chain is established and awareness about the inputs is there, then control over the output of the production process can be achieved.

COLOUR COMMUNICATION

Finally we should note another important element that is often overlooked in seeking to improve the sustainability of the textile and clothing supply chain and that is accurate and timely colour communication.

We will return to this subject later in the series but for now a few tips for improving colour communication:

- Communicate your colour accurately using physical and/or digital colour standards. Accurate communication of the colour you require will reduce lab dip rejection rates and result in better right first time bulk dyeing.
- Communicate electronically when feasible: avoid wasting time due to delayed communication
- Do not set unachievable colour standards- avoid wasting dye and chemicals in trying to achieve heavy depths or brilliant shades on certain fibres.

CONCLUSION

Sustainable design, informed selection of dyes and chemicals, accurate colour communication, and controlled coloration using Best Available Technology are key elements in raising the standard of sustainability in the textile supply chain.

CREATING A GLOBAL VISION FOR SUSTAINABLE TEXTILES

Certification, such as eco-labels, plays a major role in giving credible assurance to retailers and end consumers that products comply with standards based on social, ecological and environmental standards. Of the 309 eco-labels identified world wide, 41 cover textiles (Ecolabelling, 2008) and some 9000 textile and clothing manufacturing companies have been certified. Organic Exchange Fibre Report (2008/09) estimated a 54 per cent increase in *cultivation* of organic cotton from the previous year, but *production* of organic cotton only 0.959 per cent of conventional cotton, *i.e*. the growth in eco-labelled textiles is not reflected in consumer demand, raising questions about the impact eco-labelled or 'sustainable' textiles.

A number of issues may impede the spread of eco-labelled textiles through the supply chain: costs and time required to achieve, use and renew the eco-label, recession and potential loss of competitive advantages. This paper will present the findings from in depth interviews examining the decision making around buying and sourcing of eco-labelled fibre, fabrics or textile products.

The seven companies located both in India and the UK, spanned the supply chain, from fibre to product: textile manufacturers, eco-parameter testing labs, Certification Company and retailer. The aim of the research was to understand and investigate the marketing strategies for sustainable textile products. Our goal was to understand how designers, manufacturers and retailers may collaborate to deliver eco-labelled textiles attractive to the end consumer and we conclude by reflecting on potential implications for the supply chain integration

ECO-LABELS

Eco-labelling is becoming a differentiating factor on a worldwide scale in retail markets for textile and apparel purchase. Consumers are becoming increasingly concerned with the adverse impacts of industrial pollution on the environment and their health, resulting mounting pressure on textile, fashion industry to adopt more eco-friendly, chemicals and manufacturing processes.

Environmental concerns raised by production systems have been recognized since the late 1960's and attempts to move towards more sustainable and environmentally friendly approaches have been through a range of regulatory measures from green taxes to strict bans.

One approach acquiring increasing importance is that of 'environmental labelling' or 'eco-labelling', which, according to Piotrowski and Kratz (2005) differ in that *environmental labelling* is broad and covers a range of labels and declarations of environmental performance and focus on consumption rather than the production of a given product; *e.g*. recyclable material while *eco-labels* are a sub-group of environmental labelling and convey environmental information about a product to the consumer and communicate that the environmental impacts are reduced over the entire life cycle of a product without specifying the production practices.

In brief an ecolabel:

- Identifies the overall environmental preferences of a product;
- Provides information on environment related product qualities;
- Are tools for consumers to identify environmentally safe product;

- Enables manufacturers to use ecofriendly raw material and ingredients;
- Is an additional product quality which can be used as a marketing tool;
- Can be issued by private or public body;
- Causes less stress on the environment
- Enables to earn premium on products.

SIGNIFICANCE OF ECO-LABELS

The eco-label has a role in the Integrated Product Policy (IPP) which aims to minimise the environmental degradation caused by any of the phases of a product's life cycle (tangible or intangible, such as service), *e.g.* manufacture, development, use or disposal (European Commission, 2008).

All phases of a product life cycle are examined with the objective of improving their environmental performance. This approach requires all participants in this process to be engaged: eg, designers, industry, marketers, retailers and consumers.

The US EPA (1994) defined the following five factors for measuring effectiveness of an eco-label, the first four of which serve to support the last:

1. Consumer awareness ®of labels
2. Consumer acceptance of labels (credibility and understanding)
3. Changes in consumer behaviour
4. Changes in manufacturer behaviour
5. Net environmental gains

TYPES OF ECO-LABELS

Eco-labels may be voluntary or mandatory. Mandatory labelling is always third party labelling (*i.e.* an independent body is required to attest to required standards having been achieved), voluntary programmers may be established by firms or business associations as well as third party. Currently, there are no eco-labels in textiles and clothing enforced by mandatory rules. Eco-labels are normally issued either by government supported or private enterprises once it has been proved that the product of the applicant has met the criteria.,

- *Government*: Blue Angel (Germany), Eco Mark (Japan), Environmental Choice (Canada), White Swan (Nordic Countries), EU, Eco-Mark (India), Green Label (Singapore)
- *Private*: Eco-tex, Oeko-Tex (textiles and clothing) (Germany). Green Seal (United States),

The criteria for granting eco-labels are mostly based on the "cradle-to-grave" approach, *i.e.* the life-cycle analysis of the product and assessment of its impact on the environment from processing of raw materials, production, distribution, consumption and maintenance, (*i.e.* washing, ironing, dry-cleaning) and finally disposal of the product.

A 'Cradle to Cradle' certification programme assesses the sustainability of product ingredients for human and environmental health, as well as their recyclability or compostability making it easier at the design stage to create ecologically-intelligent products through choosing materials that meet key sustainability criteria for material health and material reutilisation (Braungart and McDonough, 2008). Differences between various eco-labelling schemes confuse public understanding of eco-labels: some are based on detailed analysis of the environmental impacts while others analyse only certain stages of the life-cycle.

Voluntary labels are classified according to International Standards Organisation (ISO) (Baumann, 2007). ISO is the world's largest non-governmental organization that develops and publishes International Standards and is a network of the national standards institutes of 163 countries. There are now eco-labelling schemes both in developed and developing countries and the ISO has classified the existing environmental labels into three typologies – Type I, II and III, specifying preferential principles and procedures for each one of them (beyond the remit of this paper to detail). Many other prominent international trade and environmental organisations deal with issues related to eco-labelling, *e.g.*: the United Nations, the World Trade Organisation through its International Trade Centre and Committee on Trade and Environment, the US Environmental Protection Agency, as well as the Organisation for Economic Co-operation and Development (Naumann, 2001).

DEVELOPING AN ECO-LABEL

This is complex and complicated but can be generalised into four broad phases:

1. Selection of a product category by a labelling board through suggestions from industry, environmentalist, consumers, and other interested parties. De Man *et al* (1997) proposed that four levels of actors function throughout the industry:
   - Primary economic–Production/consumption decision-makers (producers, importers, consumers).
   - Secondary economic–influence the decision making of primary actors.
   - Governmental and administrative–set the framework for the actors.
   - Others–try to influence the behaviour of all actors to improve status quo.
2. Life-cycle analysis to assess environmental impact of products in chosen category and examine the material and energy inputs for manufacture and use of a product and the solid, liquid, and gaseous waste generated at each stage of lifecycle, *e.g.* raw material, production, distribution, packing use and disposal.

3. Criteria and thresholds for the award of an eco-label set taking into consideration technical feasibility and environmental impacts in different media like air, water and soil against one another. Different eco-labels have differing methodologies, *e.g.*, Oeko-tex 100 examines harmful residues on the product, while GOTS tends to look environmental as well as residual parameters.
4. The product category and criteria is reviewed and refined. Interested parties including industry and environmental and consumer groups are asked for their inputs, although they are often already included much earlier on in the process.

ACQUIRING AN ECO-LABEL

Atilgan, (2007) identified fourteen steps to obtaining the EU Flower eco-label for a Turkish textile manufacturing firm In brief, the factory administration decides to apply for eco-label and use it. At the performance control stage, the selected ecolabelling organisation's prohibited chemicals are identified. Detailed monitoring of the chemicals and quality control system is necessary at each stage of production to prove that the products introduced to the certificating institute are appropriate.

At this stage, honesty is necessary because the institute is authorised to make tests any time it likes. Failing in these tests results is cancellation of the eco-label certificate (Atilgan, 2007). As well as the time and complexity of obtaining the eco-label, the most significant issue is that of cost, which, for the EU eco-label is set at 0.15 per cent of the annual turnover of the eco-labelled product, costing up to □1,300 for registration (*i.e.* to apply for the label), □25,000 per year for the use of the label, with a reduction of 25 per cent for SMEs (buyusa.gov 2009, Rubik and Frankl, 2005).

Atilgan (2007) indicated that the costs of using eco-labelled production made the finished product between 12-15 per cent more expensive to make, depressing interest in their use by manufacturers and retailers. The costs become even greater down the chain. A UK corporate wear supplier estimated the costs of using eco-labelled fabrics (such as Teijin fabrics, from EcoCircle closed loop system of fibre processing) as placing a 57 per cent premium on their final product (Sinha and Hussey, 2009).

ECO-LABELS AND CERTIFICATION

Regardless of the costs/benefits, Atilgan (2007) urged the Turkish government and industry to become engaged with eco-labelling as he felt that the next area of purchase and trading selection appeared to be based on the criteria set by the eco-labelling bodies. Furthermore he suggested that these costs may be mitigated through the use of smaller amounts of high quality products, optimising the production techniques *e.g.* by controlling all recipes and procedures, and identifying problem areas.

Given the premiums on using eco-labelled textiles, commercial buyers of such products require assurances that goods comply with standards set by the eco-label organisation. This is particularly relevant for organic fibres where the market has flooded by products described in vague terms such as 'Green' and 'Eco' (Rundgren, 1999). To differentiate between sustainable and traditional textiles, third party standards are desirable.

The unverified, market-based self-labelling (without outside monitoring) approach for textile ecolabelling, invites fraud due to lack of third-party verification (Moore *et al.*, 2009). Retailers and consumers are starting to demand labels backed by solid third-party certifications to give confidence to all the marketing claims. Commercial buyers therefore may require certification of the product. Like ISO Type I labels, certification schemes are voluntary and provide information on the environmental impacts of a company's production methods and processes (PPMs), ie, impacts of the *entire* activity not just a particular *product* (Rotherham, 1999). Certification provides a comprehensive system for ensuring that certain standards of organic production and processing are met.

The system includes:

- Developing rules or standards (standard setting).
- Verifying and evaluating performance against those standards (inspection).
- Recognizing procedures which successfully meet the standards (certification).

SCOPE AND TRANSACTION CERTIFICATES

Most companies only look apply for certification if it is required by the customer. From the literature reviewed to date, shows the flow of the certification across the supply chain. There are two kinds of certificates: Scope - issued to the selling company and stating its name, the products and production facility inspected and certified in accordance with standards. Transaction: only issued (to the seller) for the sale of products if the applicant has the scope certificate. For example, a retail buyer may require a scope certificate from chosen garment manufacturer to prove that their manufacturing facility is certified and products are being produced in accordance with standards. If the manufacturer is not certified, they contact the certification body for the certification. If the garment manufacturer does not manufacture textiles, then the retail buyer would need to source the required certified raw material to process into fabric. Then they will contact the factories from where they wish to buy the fabric, however that facility must also be certified and to prove it, processor need to provide scope certificate to garment manufacturer, if there is no certificate then they need to contact the certification body to certify their facility. If the ordered garment has embroidery and printing, then either the garment manufacturer should acquire certification for those facilities under their scope, or suggest to the embroiderer or printer to acquire certification themselves through using certified threads, dyes or chemicals.

To obtain a certificate can take between 30-45 days, depending on the standards, factory condition, number of factories under one application understanding of standards by applicant, and changes within the factory with respect to compliance with standards. This process continues 'backwards' along the supply chain till supplying of certified (in this case organic) cotton to spinner.

CERTIFICATION FLOW

If all the points are certified the process would take around a minimum of 30-45 working days to issue a certificate from certification body. If during inspection any non-compliance occurs the process would be lengthened as each certification takes about 30-45 days. A further complication is that the entire chain requires consistency in certification bodies used: *i.e.*, if the garment manufacturer is required to produce to GOTS standards, the processors must also have GOTS certification-a different certification will not suffice as they each have their own analysis processes for certification.

In brief, certification is a complex, time consuming and costly process however it creates the transparency within supply chain. For each step in the supply of a product for the chain, the transaction certificate provides transparency within the supply chain and it gives assurance to buyer about the compliance of products with standards. The certification body issues the transaction certificate for each application made by their certified clients and is delivered to the buyer to give assurance and to confirm the product is manufactured accordance with standards.

- Demand for Factory certification
- Receiving certified product from supplier
- Contacting certification body for certification
- Contacting ecological testing laboratory for required test reports

COST AND BENEFITS OF CERTIFICATION

An important factor in the success of any eco-labels is its ability to cover its certification costs and therefore stay in business.

According to EPA (1998), the ease with which programs will be able to cover costs varies depending on two questions:

- Can the programme charge enough in application, testing, audit and other fees to cover its costs; and
- Can the programme subsidize its environmental labelling activities from other programme activities?

Cost of certification can be high in relation to the value of the product and thus can become prohibitive. This is especially true for textiles because of the number of process are involved from production to consumer. It may be more expensive for companies in developing countries to obtain labels and certifications, "due to factors such as the lack of existing management structures (EMS), the novelty of EMS, insufficient infrastructure, and high auditing costs if companies have to rely on international consultants and certification companies" (Rotherham, 1999).

In addition to capital costs, the absence of necessary knowledge and skills and a lack of mutual recognition between different national programs can further disadvantage some countries (UNCTAD 1997).

ECO-LABELLED SUSTAINABLE TEXTILE PRODUCTS:

'Sustainability' is the ability to maintain an activity indefinitely over time. A sustainable activity is therefore one that does not exhaust the resources on which it depends. The concept of sustainability gained worldwide recognition following a report in 1987 by the World Commission on Environment and Development (WCED). This report entitled *Our Common Future*, defined sustainability development as (Performance Apparel Market, 2009) –

"Development that meets the needs of the present without compromising the ability of future generations to meet their own needs."

Based on above definition, it's possible to define the term 'sustainable textile product' (STP) as –

"a product which is manufactured with the help of services and related products in responds to basic need and bring a better quality of the life with use of minimum amount of natural resources and toxic chemicals during the production, minimum waste emission to the environment over the life cycle of product keeping environmental and social factors in mind throughout the supply chain".

Sustainable textile or apparel are, therefore:

- Safe for human and physical environment;
- Made from renewable materials;
- Produced while making the most efficient use of resources such as water and energy;
- Manufactured by people employed in decent working environment;
- Capable of being washed at low temperature using environmentally friendly laundering agents; and
- Capable of being returned safely to the environment at the end of their useful life (Performance Apparel Market, 2009).

CHALLENGES FOR ECO-LABELLING

There are many challenges for the eco-labelling, the most serious of which are: misleading or fraudulent to uninformative claims, unfair competition and protectionism and lack of stringency or standardisation in the process or mechanisms of eco-labelling.

The objective of certification is to gain access to the market for environmentally sustainable products (Rundgren, 1999), and the certification process should help as data collected in the process of certification can be very useful for market planning as well as for extension and research, moreover, improves the 'image of product and increases its credibility and visibility (Rundgren, 1999), Auriol and Schilizzi's (2003) studies have shown that the costlier the certification process, the fewer firms able to afford certification, ie, cost becomes a major factor in deciding market structure, potentially leading to monopoly and ultimately to no certification at all.

Influencing factors on consumers' willingness to buy environmentally friendly products have been identified and categorised as: demographics, knowledge, values, attitudes and behaviour (Laroche *et al.*, 2001). However, price has been found to be one of the most decisive factors in determining when consumers actually purchase apparel products.

Consumers' willingness to pay and purchase cloths made from sustainable raw material like organic cotton is a complex issue (Gam *et al.*, 2010). Empirical testing has shown, however, that many consumers are willing to pay a premium for eco-labeled products (Imkamp, 2000; Loureiro *et al.*, 2002; Makatouni, 2002; Moon *et al.*, 2002) and that they do purchase such products (Lathrop and Centner, 1998; Teisl *et al.*, 2002). According to Gam *et al.*, (2010) study in one of US, only 35 per cent (27 out of 84) were willing to pay more for OCC and only 10.7 per cent (9 out of 84) were willing to accept more than a 10 per cent increase in price for OCC. In contrast, they found that 52 per cent of survey participants would pay a 50 per cent price premium and 25 per cent of the participants would pay 100 per cent more for an OCC over conventional cotton clothing. From above results it's clear that, willingness to pay for STP like product made from organic cotton varies place to place, country to country. Differences in testing and certification methods have created difficulties in the application of an eco-label to a particular product category. For example, should the label represent an overall assessment of a product's environmental burden over its entire life cycle, or some subset of it? What techniques can be used to measure environmental impact? Who determines what specific environmental impacts are the most important? And what criteria are appropriate in rating impacts? Moreover, the consumer is unable to verify the claims made by the eco-label.

An analysis of ecological labelling process by Lavallee and Plouffe (2004) concluded that 'cradle-tograve' analysis for ecolabeled products and services is not always, in fact, respected, and that at the present time ecolabel delivery criteria are not sufficiently stringent or standardised leading to confusion in the marketplace, making it difficult for companies to identify stakeholder preferences and for justified environmental claims to be considered credible (Rotherham, 1999; EPA 1998).

THE RESEARCH QUESTION AND METHODOLOGY:

The research undertook to understand the issues within and across the textile supply chain that come to bear upon the growth of eco-labelled sustainable textiles products. While the fear of losing market share is a motivating factor, it should be stressed that market impact of eco-labelled textiles products are only one indication of an ecolabelling programme's success.

The effectiveness of an eco-label ultimately depends on the extent to which consumers perceive, recognise and act on the information it conveys. The next section describes the research methodology adopted and presents some of the most pertinent results. Multiple in-depth interviews were conducted using semi-structured interview schedules designed as open ended questions. The decision was made to focus the study on organic cotton (one of a range of sustainable fibres) as this (and its products) is produced in large quantity, consumers are much more aware of it, retailers as well as consumers have accepted it and year by year demand is growing throughout world. 9 in-depth interviews were conducted with company employees who ranged from director of development, director of Marketing, Company Owner/Director, etc. As indicated, the companies were selected on the basis that they had at least one following production facility in order to complete as much as possible the supply chain: spinning, weaving, knitting, wet processing, stitching/garmenting, retailing, testing, authorization of certification and authorization to issue eco-label. Companies interviewed are listed in table 1; the majority of the companies were in India as this is a centre of organics cotton production.

RESULTS

As mentioned earlier, the research tried to understand the issues within and across the textile supply chain that come to bear upon the growth of eco-labelled sustainable textiles products in particular and in developing a sustainable textiles industry in general. From the interviews, the following issues arose:

SUSTAINABLE TEXTILES PRODUCTS (STP) ARE NEEDED

All companies interviewed agreed that STPs are needed. According to textile manufactures companies A and B, STP is a holistic approach and it can be achieved through recycle, reduce and reuse processes. They felt it very important to note that all naturally grown products are not organic or sustainable; for example, organic cotton. All naturally grown cotton is not organic; it might be genetically modified organic cotton. Also, there is no assurance that the land doesn't have any traces of harmful fertiliser, pesticides. According to dyes and chemicals manufacturers companies D and F, STP is a mindset (company D) and its "the product which has manufactured by taking care of all the three elements of sustainability that is: social, economic and environmental sustainability and product design is fashionable to sell and sustain in the market" (company F). STP's are needed to take care of our eco-system, so that this planet can 'sustain' the lives and livelihood 6 billion inhabitants.

From certifiers (company G), STP's are those products which are manufactured from fibre cultivated by natural or organic method considering social and environmental impact, understanding soil fertility and animal welfare. Further, Mr A (Company H) added that "producing sustainable textile is one of the ways to overcome the Global warming" and Ms. AB (company I) suggested product made from 'organic cotton' are the best example of STP.

'WHY' AND 'HOW' TO BECOME SUSTAINABLE TEXTILE MANUFACTURER

A number of methods are available to enter the sustainability arena; those raised by the interviewed companies included the following:

TRANSPARENCY AND EXTENT TO WHICH THE COMPANIES HAD TAKEN UP ECO-LABELLING

All the companies interviewed were either certified or were in the process of receiving certification, ie there was felt to be a general endorsement of the idea of eco-labelling.

BOLSTER THE CLAIMS WITH INDEPENDENT VERIFICATION

Third-party verification of environmental credentials can often bring legitimacy to sustainability. Many of the most successful eco-labels are those that have been backed by issues-led organisations, for example GOTS certification for textile products made from organic cotton. Third-party verification can range in scope from qualitative assurance of general claims to detailed verification of all stages of a full life-cycle product assessment.

Given the generally low levels of consumer trust in big business, some degree of external verification is an essential component of any credible environmental claim. According to Mr. H (company G) and Mr. A (company H), the claims made by the manufacturer or retailers are cross checked by certification of responsible eco-labelling body by testing the goods which can be picked from market store to confirm whether claims are right or wrong.

In case of dyes and chemical, customer requires proof from third party initiating business. Therefore many companies use the test reports as a marketing tool to prove the product and company integrity. Within this study, it was the analysis, researcher has observed that, all the segments of supply chain apply for certification or verification as a buyer requirement.

EDUCATE, ENABLE, AND ENCOURAGE

Educating, enabling, and encouraging people to act towards sustainability is key for the success of any eco-label and STP as consumers' usage and disposal patterns liberate CO2 and so there should be programmes to educate them. Methods to do this range from placing trust worthy eco-label with required info on it, through various media, and through regulations.

Once the consumer gets educated then may be encouraged to prefer buying more sustainable products by linking them with promotions and reward schemes (carefully and consistently in accordance with principles of sustainable consumption). Product information can now be shared via many more 'touch points': at point of sale; in retailer magazines, leaflets and web sites; through road shows, help lines and education packs. Enlisting employees to promote sustainability is another method as employees are key players of any manufacturing facility. Investing in the skills of employees is part of a sustainable and responsible human resource management (Brito *et al.*, 2008). To educate their employees, company A celebrates days like Green Day, Earth day in factory to increase the awareness of sustainability.

Company B have organised seminars for their employees, their suppliers, buyers, contractors etc before starting production of organic cotton where they have invited expertise from industry on subjects like eco-friendly dyes and chemicals, GOTS, ethical production practices etc.

REDUCE, REUSE, AND RECYCLE:

To become sustainable, everyone, from manufacturer to consumer should think of 'Reduce', 'Reuse' and 'Recycle'. Textile waste in landfill contributes to the formation of leachate as it decomposes, (which has the potential to contaminate groundwater), methane gas (a major cause of greenhouse gases contributing to global warming) and ammonia (highly toxic for land, water and air) (Productivity Commission 2006).

The companies interviewed engaged currently in the following activities to try to reduce waste and reuse material:

Company A:

- By utilizing solar energy company is running around 20 computers within factory,
- From cutting waste, manufactured 5000 bags and sold at local super market subsidiary price
- Currently working on yarn manufactured from solar energy,
- Reusing treated effluent water for gardening and washing purpose (specifically washing of printing screens).

Company B:

- Have reverse osmosis (RO) plant for purification of processing effluent, after purification of that water it utilised for washing, gardening etc.
- Have started recycling of old paper cones (used in spinning for packing) and making new ones.

Company I:

- Offering shopping bags to their customers made of potato starch *i.e.* 100 per cent biodegradable.

DISCUSSION AND CONCLUSION

The proliferation of voluntary certification and labelling schemes for environmentally and socially responsible production is often seen as a driven by companies and consumers. Consumers are heavily involved in environmental pollution because of their buying behaviour and consumption of textiles and significant associations were found between environmental shopping attitudes and behaviour and willingness to pay more for organic cotton products and observed that consumers with a greater environmental awareness demand more environmentally friendly merchandise (Fraj and Martinez, 2006).

Eco-labels are not simple to understand as they appear and so may not be as appropriate marketing tools as suggested by the government policies. For example, GOTS and OE standards labelling guide are both standards applicable for products made from organic cotton on which retailers, manufacturers can use respective logo on their tags.

Under both the standards it's mandatory to mention the percentage of organic cotton on the label and if the product is made from 100 per cent organically cotton, manufacturer or retailer can use the statement "Organic" and "Made with 100 per cent organically grown cotton" respectively. However, GOTS is based on social, technical and environmental areas while Organic Exchange 100 Standard (OE 100) is for tracking and documenting the purchase, handling, and use of 100 per cent certified organically farmed cotton fibres (or organicinconversion cotton fibre) in yarns, fabrics and finished goods.

The significance of this difference that, while purchasing the product a consumer will check only the organic content and assume that product is eco-friendly or sustainable. However, this is not a completely accurate picture as an OE standard does not look into social or harmful dyes and chemicals or about the environment related issues and an OE 100 logoed garment may be made from 100 per cent organically grown cotton but then finished with harmful dyes and chemicals, printed with non-eco-friendly printing technique like solvent based printing.

Therefore a large problem in marketing communication is using eco-labels is the lack of common definition or general understanding exists for what constitutes environmentally friendly clothing and eco-labelling. Eco-labels backed by solid third party certification give confidence to consumers about the genuineness of product. Certification bodies are the key player for the growth of eco-labelled STP who confirms the product, process and manufactures integrity with respect to sustainability. In this study, across the whole supply chain, most of the people had positive thought about requirement of certification.

However, cost of certification can be high in relation to the value of the product and thus can become prohibitive. This is especially true for textiles because of the number of process are involved from production to consumer. To overcome the certification and labelling cost, Ibanez and Grolleau (2007) suggested "carrot" approach - the firm who preserve the environment appreciate them by subsidizing a recognized labelling and "stick" approach - increase of labelling cost for polluting firms by enforcing stricter labelling guidelines and severe punishment in case of deceptive use of environmental claims.

According to Getz and Shreck (2006), despite much analysis of third-party certification, little is known about how certification is enabled or enacted at the point of production. The insights of those few who have explored some of the political and social effects of certification at the point of production are worthy of further examination. In this study researcher has analysed the 'how' certification process works within the organic textile-clothing industry and 'why' it is required. Having reviewed the process of certification and how individual actors of supply chain get involved into this very complex, time consuming and sometime costly process of certification. Therefore we suggest that manufacturers take a vertically integrated approach through networking or developing two or three processing facilities to overcome cost and time of certification, reduce handling, transportation CO2 emission and time to retailers' shelves. Companies such as M&S, H&M, Zara, who are bringing STP's to the mass market must be operating in this manner.

Moreover we suggest, a fourth to the traditional three pillars of sustainable development; "technical" by which the first three may be achieved.

- Demand for Factory certification
- Receiving certified product from supplier
- Contacting certification body for certification
- Contacting ecological testing laboratory for required test reports

SD pillars Why How

- Economic Competitiveness
- Process and product innovation; Process and product substitution.
- Environmental Clean out puts Reduce, Reuse and Recycle
- Social Social fairness Better human resource management
- Technical Eco-friendly inputs Use of certified products

RECOMMENDATIONS FOR FURTHER RESEARCH AND INDUSTRY

There are numerous possibilities for future research in eco-labelled STP. For this research manufactures were from India, results may vary from country to country.

However, based on general understanding, the study results indicate the following areas of further research:

- Most of the research is carried out to identify the consumers' willingness to pay for sustainable, organic or fair tread products. However, research has been not carried out about

  - How much consumer knows about eco-labels, and eco-textile standards and
  - The shopping behaviour and attitudes of STP consumers have not yet been analysed through actual purchase data.
- Investigate the health claims dictated by the lack of direct categorical data available on environmental benefits claims versus health benefits claims on labels (Nimon *et al.*, 1999). Are there are any serious health benefits of STP especially Organic Cotton to human being?
- Why are eco-standards not mandatory throughout the textiles industry?

LIMITATIONS TO THE RESEARCH

This study's limitations need to be taken into account when considering its contributions to theory and findings. This being a one year period of study, time and cost were main limitations and affected the sampling, since it was impossible to visit entire population due to budget restrictions, large dispersal of locations.

It was also not possible to include either the logistics sector (also responsible for CO2 emission); government authorities' whose views may put some light on government policies and approach towards eco-labelled STP or the manufacturers who are not certified by any of the standards to understand why they are not willing to get certify.

The results have not been validated, again due to lack of time. Bias may arisen, some participants tend to express views that are consistent with organic cotton, GOTS, OE standards only and try not to present themselves negatively. In addition, the interviewees may have also been unable to disclose some information due to the privacy issues with various companies. Information collected may be prone to some inaccuracy as a result of less than accurate recall, lack of information, or discomfort with self-disclosure.

MARKET METRICS

UNITED STATES

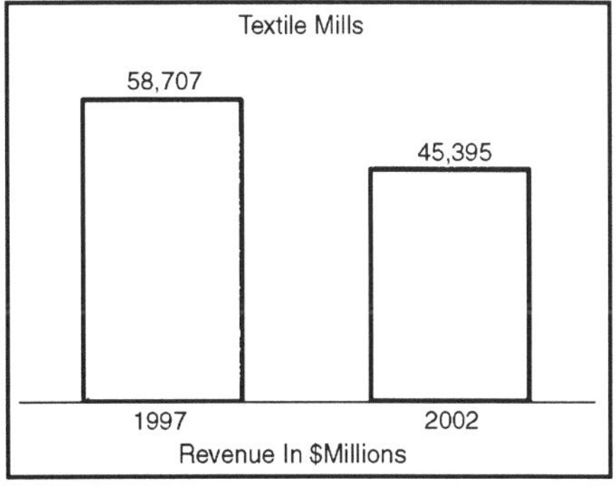

The textile industry is currently more than US $400 billion and continues to expand. The globalization of the textile trade has increased the outsourcing of production.

A variety of fabrics are used worldwide in different applications such as apparel, household textiles and furnishings, medical equipment, industrial and technical products. Recent studies have highlighted that fabric weaving alone consumes around 28 million tons of fibre annually. It is predicted that global production will grow by 25 per cent between 2002 and 2010, to reach more than 35 million tons and Asia is one of the key regions for growth.

The United States ships raw materials to other countries in which the products are made. The cost of labour is cheaper and provides retail companies the ability to buy and sell at lower costs. This increases competitiveness and has led to several textile companies having to shut down operations.

At present, there is a minor trade war developing between the United States and China in regards to textiles and apparel. Chinese exports of textiles to the United States increased by 320 per cent between January 2002 and November 2003 while prices of Chinese textiles and apparel declined by 55 per cent. The growth in imports of textiles to U.S. during 1995 to 2002 was about 9 per cent annually in dollar terms.

INDUSTRY PLAYERS

These are major players in this market, but not an exhaustive list of all key firms. Revenues, Net Income and Market Capitalization are expressed in US$ Millions.

Ticker	Company	Location	Revenue	Net Income	Profit Margin%	Market Cap
VFC	V F CP	Greensboro, North Carolina	6860	590	7.81%	8040
LIZ	LIZ Claiborne	New York, New York	4950	149	2.75%	2410
HBI	Hanesbrands	Winston-Salem, North Carolina	2250	74	3.28%	2770
RL	Ralph Lauren Polo CP	New York, New York	4540	387	8.50%	6870
ZQK	Quiksilver	Huntington Beach, Califormia	2362	93	3.94%	1280

Research and development and design are important in industrial textiles where the material technology provides competitive advantage. In industrial countries, notably the United States, an increasing share of the textile sector produces household appliances and other industrial fabrics such as for the furniture and car industries.

This is a more R&D intensive segment of the industry and subject to less frequent changes in patterns, material and colours. Only about a third of US textile production was used for clothing in the late 1990s.

- Mohawk Industries, Inc., is the market leader both in terms of market capitalization and annual sales in 2006. It engages in the production and sale of floor covering products for residential and commercial applications in the United States and Europe. It operates in three segments: Mohawk, Dal-Tile, and Unilin. The Mohawk segment designs, manufactures, and distributes floor covering products, including carpets, rugs, ceramic tiles, hardwood, resilient, and laminates. It markets its products through independent floor covering retailers, home centers, mass merchandisers, department stores, commercial dealers, and commercial end users, as well as through private labeling programs.

- Albany International Corp. produces paper machine clothing and doors worldwide. It operates in three segments: Paper Machine Clothing, Applied Technologies, and Albany Door Systems. Paper Machine Clothing segment includes fabrics and belts used in the manufacture of paper and paperboard. This segment offers forming, pressing, and dryer fabrics, and process belts.
- Polymer Group, Inc. manufactures and markets non-woven and oriented polyolefin products. Its Nonwovens division provides non-woven materials, which are used in diapers, training pants, feminine sanitary protection, adult incontinence, baby wipes, and consumer wiping products. Its products are used in medical applications, such as wound care sponges and dressings, and disposable surgical packs, as well as for apparel, including operating room gowns and drapes, face masks, and shoe covers; and in industrial applications.
- Interface, Inc. engages in the design, production, and sale of modular carpet products in Americas, Europe, and Asia-Pacific. It also manufactures broadloom carpet, panel fabrics, and upholstery fabrics for the commercial interiors industry.
- International Textile Group, Inc., together with its subsidiaries, operates as a textile manufacturer. The company produces automotive airbag fabrics and airbag cushions for automotive airbag modules. International Textile Group also manufactures and sells various technical and value added fabrics for use in fire service apparel, ballistics materials, filtration, military fabrics, and outdoor awnings and covers. The company also produces denim, cotton, synthetic and worsted apparel fabrics for use in bottom garments, such as pants, skirts, and shorts. It manufactures fabrics for military dress uniforms and battle dress uniforms for the U.S. Government and government contractors.
- Xerium Technologies, Inc. manufactures and supplies consumable products used in the production of paper clothing and roll covers.
- The Dixie Group, Inc. engages in the manufacture, marketing, and sale of carpets and rugs to residential and commercial customers in the United States.

TRENDS AND RECENT DEVELOPMENTS

To sharpen their competitive edge, many U.S. textile companies are,

- Restructuring their operations through mergers, acquisitions, and divestitures,
- Concentrating on more profitable niche markets and product diversification,
- Improving productivity through new capital investment,
- Building strategic partnerships with retail and apparel companies,
- Using quick response and other communication-oriented techniques to enhance their overall flexibility and competitiveness.

U.S. textile companies have been acting to improve their competitive positions through reorganization, restructuring, and downsizing. By merging, textile companies gain market share and access to capital. Through acquisitions and vertical integration, textile companies have been able to achieve economies of scale and expand the range of their services.

Another trend is for U.S. mills and converters to move slowly away from staple fabrics for the apparel industry and towards niche products and diversification into multiple sewn products markets. Technical fibres for sportswear and active-wear have experienced tremendous growth. Other profitable growth opportunities in non-apparel fabric markets include automotive, medical, fire safety, and telecommunications (fibre-optics).

The U.S. textile industry is a technologically advanced industry, achieving such advancement through restructuring and investment in modern machinery. Spinning, weaving, and knitting processes are highly modernized in many companies, and this has allowed them to reduce the share of labour in production costs and increase quality and speed.

There has been a substantial job loss in the U.S. textile industry, a trend that has been ongoing for almost 30 years. This job loss can be attributed partly to imports, but like many other industries in the United States, the textile industry has been able to increase its production through new technologies while decreasing the number of employees.

Overall textile production has increased more than 20 per cent in fifteen years even though industry employment has dropped over 13 per cent. The textile industry continues to invest in machinery and technology, spending upward of $2 billion a year to maintain modern manufacturing facilities. Continued investment in modern technology, including information technology, is very important. Textile production is very capital intensive, and up-to-date technology is essential to meet the increasingly rigorous demand for high-quality products.

Proximity to major markets has assumed increasing economic significance and tariffs are restraining trade due to the fact that products cross borders several times. Developing countries are catching up with China in terms of unit labour costs in the textile and clothing sector. Future trends include multifunctional textiles, intelligent textiles, eco-textiles, e-textiles and customised textiles in the textile-apparel industry. Nano-Tex is the company manufacturing "enhanced textiles" which are stain resistant transparent fabrics manufactured using nanotechnology.

4

Industry of Synthetic Fibers

There has been tremendous growth of synthetic fibre industry in India count Hilaire chardon was the first to make a fibre similar to silk. He made it from nitrocellu lose by dissolving it in ether. The first truly man-synthesized fibre, nylon was produced in 1938. All man-made fibres have some processes in common.

They have been produced f.rom non-fibrous materials or if fibrous material were used originally, then they have lost their fibrous structure during processing. This is done by forcing the solutions through spinnerets.

These fibres are then permitted to harden during a specific. time and then are wound on bobbins or cones or they are deposited in "pots" as cakes of yarn. Man-made fibres are divided into two major classifications-non-thermoplastic and thermoplastic. The non-thermoplastic groups of man-made fibres have several origins, *e.g.*, cellulosic, alginates, minerals and protein base fibres.

Barring for the mineral. based fibres, these non-thermo-plastic fibres can be cared for as cotton, silk, wool whichever they similar most in their physical and chemical reactions. Heat will not melt them but they scorch easily at high temperatures. They are soft, absorbent, comfortable to wear, do not discharge static electricity and are mothproof.

Like all other man-made fibres, under the microscope they appear as smooth rods, with black specks, as if they are delustered. None of these fibres can be positively identified in longitudinal sections by the microscope.

The first thermoplastic fibre to be made was acetate followed by nylon. Now apart from the acetates and nylons, there are acrylics, modacrylics, nefrils, olefins etc. These thermoplastic fibres react very fastly to heat. They soften and become pliable. This softening point varies with different fibres.

When these fibres are at the softened stage, they can be shaped, pleated or embossed. These fibres, like other man-made fibres, must be stretched to orient the molecules in order to increase fibre strength, their tenacity, or to give them dimensional stability. Nylon is generally heat set to stabilize the twist of yarn, to remove shrinkage, to increase wrinkle resistance, to set the grain in the fabrics, or to make washable pleats. These thermoplastic fibres have a number of properties in common. Their hygroscopicity is low.

Therefore they are uncomfortable in hot humid weather. They are washed easily and dry quickly.

Water soluble stains are removed easily but not grease or oil. Majority of these fibres accommodate charges of static electricity. They are mostly wrinkle resistant. A microscopic examination reveals that they have the same longitudinal appearance as other man-made fibres. But their cross-sections vary and they can be identified with these.

VARIOUS FIBRES OF NON-THERMOPLASTIC NATURE

The Rayons

Rayon is an artificial, synthetic fibre made from cellulose.

Existence

The rayon fabrics were at first called 'artificial silks', as these resembled the natural silk in appearance. As recently as 1924, the name 'rayon' was provided to these fabrics. The rayons produced then were very lustrous, and, therefore, the name which means 'reflecting the sun's rays' is considered very suitable. Although rayon fabrics have been commercially available for little more than fifty years, the idea of making such fabrics dated back to 1664 A. D. Robert Hook, an English Naturalist, prophesied in the year that a way would be found to make an artificial fibre which will similar natural silk.

Then two centuries later, in 1884 Count Hilaire Chardonnet made a fibre which resembled silk, from nitrocellulose by dissolving it in ether, and produced it on a commercial scale. Count Chardonnet is, therefore, considered the father of rayon fibre and the rayon industry.

But the production of rayons did not make much development. In the startings of this century, U.K., Switzerland, Belgium, Austria, Germany and U.S.A. started making rayons. Since then much experimental and research work has been done to improve the process, as well as the quality of the yarn, in the course of which many processes were discovered and perfected.

These are called as:

- The viscose process,
- The cuprammonium process
- The cellulose acetate process.

Now all these processes are more in use than the one discovered by Chardonnet which is much more expensive. India has been importing rayon fabrics as well as rayon yarn. Since 1942, however, the Board of Scientific and Industrial Research has been making attempts for starting factories for making rayon yarn and many a plants have been set up since then. Bamboo, and

Bagasse-cellulose from sugarcane a by-product of the sugar industry is used for making rayons in India.

Rayons Making Principles

Rayon, is made from cellulose by anyone of the former particular processes, but the important steps in each of these are:

- To treat cellulose chemically for making and rendering it a liquid,
- to force the liquid through fine holes,
- to change the liquid stream into solid cellulose filaments.

Process of Nitrocellulose

In this, cotton is reduced to nitrocellulose by treating cotton with sulphuric and nitric acids. This gives an inflammable material. This material is dissolved in ether and the fluid thus formed is compelled through tiny holes, into air. The ether evaporates and a fine thread is obtained, which is treated with sodium hydrosulphide to render it non-inflammable.

Process of Viscose

This process was discovered in 1892, but the yarn was commercially produced a few years later. In this process spruce. chips are utilised. These are first bleached, and then steeped in caustic soda to form alkali cellulose. It is treated with carbon bisulphide to form cellulose xanthate and is dissolved in dilute caustic soda solution. This is filtered and kept for ageing until a thick fluid is formed, which is known as viscose.

This fluid is forced through fine jets into a coagulating. Solution of dilute sulphuric acid, which regenerates the cellulose into a continuous fibre. This process is mainly used in the manufacture of rayons. If a dull appearance is desired, a small proportion of a whiter opaque pigment (generally titanium dioxide) is added to the viscose solution.

Process of Cuprammonium

This process has been utilised since 1897. In this process, cotton linters are used. These are first boiled in soda and soda-ash, and then are bleached with chlorint.

These are dissolved in a solution of copper sulphate and ammonium hydroxide. The liquid is left for ageing or ripening. This is then passed through fine jets into a solution of dilute acid when it is changed into regenerated cellulose filament. This filament is stretched to form a fine thread.

Process of Cellulose Acetate

This process has been commercially developed since 1918 although the process was discovered half a century earlier in 1869.

Rayon Yarn Spinning

The cellulose solution is passed through the spinrerets into a coagulating medium, in which the fibres get hardened. Many fibres are twisted together to form a yern. The yarn then is wound onspools; winding and rewinding is repeated using two spools and rewinding is repeated using two spools and each time giving a twist to the yarn. Then, finally, the yarn is would into skeins.

Rayon Its Features

Effect of Moisture and Friction

Regenerated rayons loose strength when wet but cellulose acetate is not as much affected in that respect by moisture. On drying, all of them regain theirstrength. Friction will weaken and spoil the lustre of the fibre, specially when wet.

Conduction of Heat

They are good conductors of heat.

Hygroscopic Moisture

Rayon fabrics absorb more moisture than cotton or linen but do not give out moisture as rerdily as those two fabrics.

Heat Effect

Heat affects them. Cellulose acetate rayon melts and fuses by the application of heat. The viscose and cuprammonium fibres are not greatly affected by heat, and the delustered effect given to these fabrics by heat can be diminished by the application of heat and moisture together as this will bring back the lustre.

Acids and Alkalies Action

All rayons are weakened even by dilute solution of acids. Cellulose acetate dissolves in acetate acid and formic acid Alkalies also have harmful effect on rayons.

Acetone Effect

Cellulose acetate dissolves in acetone, but not the other rayon.

Dyes Affinition

The viscose and cuprammomum fibres have great affinity for dyes used for cotton and linen, but the cellulose acetate rayon does not react to the same dyes. Rayons generally take dye-stuffs readily.

Bleaches

Action reducing bleaches may be used in cold, dilute solutions. Oxidizing bleaches will damage the fibre if these are not used with great care.

Rayon Lasting

Spun rayon made from tiny fibres is not so strong, but rayon fabrics made from long fibres are very strong, though net quite as durable as pure silk, cotton or linen. They are, however, generally more durable than weighted silk.

Elasticity

Rayon does not have natural elasticity. and, therefore, rayon clothes frail at the elbows, knees and the seams. Rayon requires greater allowance for seams. It is also important to have even tension and length of the stitches to avoid slipping. There are a number of non-thermoplastic fibres besides nylon, the foreign market. Some of them are avril, zantrel, qulin, Fortisan, Cornal, Tikel. The processes utilised for the new fibres have not been published, but their preparation are known. Compared to regular rayons they have improved dimension stability, better wet stability and they have lower power of elongation.

Fibres of Thermoplastic Nature

Nylon

It was the fibre first to be synthesized from materials none of which had previously been fibrous in nature.

Origin. The history of nylon dates back to 1928, when Dr. Stine the Chemical Director of du Pont Company prevailed upon the Company to begin research work relating to new fibres. The research continued for some time and experiments were made on cellulose derivations, particularly the esters arid new types of esters, and also on certain nitrogen comprising derivatives of cellutose.

These experiments were not successful Dr. Wallace H. Cenothen, at the same time made a study of poly-condensation whereby linear-polymers are produced, and this eventually led to the invention of nylon. In 1938, du Pont Company begen a plant for nylon production. This fibre was known as fibre 66. Later, the name nylon was given to it.

Nylon Making Process

Although it is often stated that nylon is made of coal, air and water, the actual synthesis is completely different. Nylon 66 is produced from an acid and a diamine, which are produced from coaltar derivaties. A mixture of two coal-tar products, a dibasic acid, adipic acid, and hexamethylene diamine containing nitrogen is heated to give a condensed product called nylon polymer. The, process is explained diagrammatically. Since each of the above compounds have six carbon atoms, the finished fibre made from this was known as Fibre 66.

The nylon polymer is me1ted and passed on chilled rollers when it comes out in the form of rolls, these are then cut into chips and stored. The chips are remelted and iltered through special filter packs. After filtering, the molten polymer is passed through the tiny holes of a metal disc called a spinneret and fine filaments are town. out into air where these get hardened.

These are then passed through a conditioner which moistens them so that a number of these filaments stick together to form a long thread which is wound on a reel. In order to increase the tensile strength and elasticity in the fibre, the filaments are drawn to about four times their original length by the application of force. In this operation tbe diameter of the fibre is reduced. Twisting of the yarn is also carried on at the same time by the use of suitable equipment for the purpose.

Nylon: Its Characteristics

It is a lustrous white transparent fibre and is both tough and pliable. Absorbency is low and this makes it a quick drying and easily washable fibre. Although it does not stain readily it tends to pick up colour, grease and soil if laundered with other garments. It is not affected by cold temperatures but looses strength and yellows at sustained high temperatures.

Pressing temperature of *24°F* is considered safe for nylon. It has good wrinkle resistance and crease recovery. It has very good abrasion resistance and is considered a very durable fibre. It is degraded by exposure to light. It is not chemically reactive to soap, alkalies or alcohol but reacts soon to acid.

Bleaches do not affect it and so are ineffectual for whitening discoloured nylon. They can be dyed and usually the colours are permanent. They are sometimes resin treated to stabilize the weave, silicone treated to improve softness or water-repellancy but because of their low absorbency they do not accept finishes readily.

Dacron

The work of Carothers paved the way for several other textile materials. Terrylene is one of these. Whinfield ard Dickson patented the method for making terrylene utilising teraphalic acid and ethylene glycol. This acid and alcohol are polymerized in vacuum at a high temperature and the polymer is extended in the shape of a riboon. The polymers (molten) are passd through spinnerettes with circular perforations.

The Pont Company purchased patent rights of this fibre in the United States, and with some changes in processing, put Dacron as a new fibre on the market in 1953. Dacron and terrylene are essentially the same fibre. The most significant properties of this fibre are its excellent resistance to wrinkling and creasing, both when dry and wet. It has good abrasion resistance, toughness, resilience, elasticity and stretch resistance.

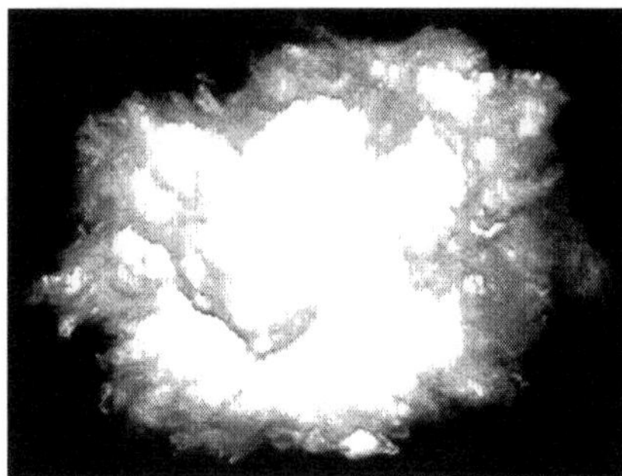

Fig. Dacron

Dacron has a property called as 'wicking'. Wicking is the ability of a fabric or fibre to pick up moisture and allow it to travel along the fibre topick up moisture and permit it to travel along the fibre, without actually absorbing it. Thus dacron is comfortable to wear inhot weather. Dacron melts and drops when exposed to fibre.

It is self-extinguishing and does not flash burn. It has better resistance to weak acids and alkalies but is disintegrated by concentrated sulphuric acid. Dry cleaning solvents and bleaches do not effect this fibre. It is not easy to dye but can be satisfactorily coloured with acetate dyes.

Natural Proteins Based Materials

- *Ardil:* The raw material for the manufacture of ardil are the proteins, arachil and conarchin which can be obtained from ground. It is a consequence of the research carried by Imperial Chemical Industry (I.C.I.).
- *Fibrolane:* In principle the manufacture of fibrolane is resemble to that of ardil, the chief difference being that casein from milk is used instead of groundnut protein. People will be surprised if they are told that the suits they are wearing are of like instead of silk.

Mineral Fibres

These minerals are brought into use in textiles fibre making; asbestos a natural fibre; glass fibre and metallics a man-made fibre.

Asbestos Fibre

It is a naturally taking place mineral fibre. It is fibrous rock and is famous for its non-burning qualities. Under the microscope the fibres appear straight, smooth and needle-shaped. It does not burn or melt but short periods of high temperatures will reduce its strength. It is absorbent and has Yicking ability. It is acid and alkali resistance and this makes it particularly desirable in the use of filters for chemicals and in industries where chemicals are utilised.

Flameproof clothing for industrial and military purposes are also made from asbestos. It is used for all kinds of fire fighting equipment and insulation for steam pipes, brakes, and hundreds of other items where non-combustibility is inevitable.

Metallics Fibres

Metallics are defined as "a manufactured fibre composed of metal, plastic-coated metal, metal-coated plastic, or a core completely covered by metal." They are yarns and not fibres, but they are considered in fibre groupings as they replace fibres for textiles purposes. Gold and silver have long been utilised in India in textile fabrics.

Aluminium and copper have a more recent origin. Metallics were used to add richness, glamour and glitte to fabrics.Modern metallic fibres, with aluminium as the basic metal, are much cheaper, softer and lighter in weight. But because they are no more rich and luxurious and look cheap this may lead to their rejection.

Glass

'Fibre' is the trade mark for the glass fibre produced by the Owens-Corning Fiberglas Corporation of America and is die only Glass fibre about which information is available. Glass fibres are inorganic polymers based on silicon rather than carbon. These fibres have not become very popular as they are subject to severe abrasion and cut each other if they rub together. They are used for curtains and drapery fabrics due to their attractive appearance, non-combustibility and are resistant to light deterioration.

They are not suitable for wearing apparel as the ends of these fibres irritate the skin, they are heavy and absorbent. Much of the gtass fibre is used for insulation in buildings and for insulating electrical and other industrial equipment.

5

Construction of Spinning Yarn

When bundles of fibres are drawing out and twisting by a process called spinning. In the starting, the yarns were spun by man with bare hands without the help of any tools and it must have been several centuries before the spindle was evolved for spinning.

Fig. Spinning

The spindle or 'TakIi' still survives for spinning wool, silk and cotton yarn. It is the simplest tool used for spinning which comprises a round disc which is attached in the centre to a thin, smooth rod about 7 inches long. The upper end of the rod has a groove or a hook which catches up the fibres. When the spinner pulls the bulk of the fibres, these are drawn out in a long strand. The spinner while simultaneously pulling the fibres, provides a twist to the spindle and lets it go. The spindle whirls round, and thus, twisting up the pulled fibres makes a continuous thread.

The thread is then woven round the rod above the disc. The disc gives the essential weight which quickens, increases and prolongs the revolving of the rod, thereby, the twisting of the fibres ensures a strong thread.

SPINNING WHEEL OR CHARKHA

Who does not know the, *'Charkha'* in India? A *Charkha* is for spinning yarn in the handloom industry in India. The yarn spun or the charkha is of different qualities. The famous 'Dacca muslins', which were uncomparable for their fineness, but alas are extinct now, were woven of a charkha yarn. The immitable soft and light pashmina of Kashmir are still woven of the charkha spin yarn.

Fig. Spinning Wheel or Charkha

In the textile industries today, electrically driven machines are employed for spinning. Many machines are used to complete the process of spinning, which of stages, such as drawing, out of the fibres in silvers, further drawing out of silvers, further reduce size and to give slight twist-roving and then spinning.

There are two methods of procedure. In one process, the action of drawing, twisting, and winding is regular and this is known as 'ring' spinning, and in the other, the draw and twisting is stopped while the twisted thread is wound up (as in the case of hand spinning) and this is known as mule spinning'.

The 'ring' spinning is a quicker process and has the benefit of reducing operating cost and increase production, but the mule spun yarn is finer, softer and of greater evenness. In spinning yarns, whether hand or machine, a difference has to be made in the yarn intended for warp and weft or filling.

The wrap yarn requires a greater amount of twists to produce strong, firm thread which is used for the foundation of the fabric and is subjected to mire strain and friction during the process of wearing.

The yarn is provided twists of a particular number of turns to the inch and are given either 5 or 2 twists. At first, regular thread or a strand of twisted fibres is made and save ral of these strands are twisted together to get the final yarn.

COUNT

The yarn is described forsize or count. A skein of yarn, 840 yards in length, is known as a hank and is the ground for determining the count. If a hank of 840 yards weighs one pound, the yarn is number one, and if two hanks weigh one pound the yarn is of count 2. Several variations are introduced in the preparation of yarn by mixing different sorts of fibres, by mixing various coloured threads or to produce texture, decoration or to add particular properties to the fabrics. The variations in yarn may be divided into two main groups:

- Novelty yarns.
- Simple.

Novelty Yarns

The construction of these yarns is of a complex nature and is varied in many ways. These yarns are usually ply yarns of different kinds of fibres or of separate colours and are irregular rather than smooth.

Single strand or yarn of various colours, sizes or fibres may be twisted together to form one complex yarn. Another kind is brought about in this type of yarn by varying the tension or speed after, intervals of certain length thus, allowing one part to loop or twist around the other. Novelty yarns are also prepared from simple yarn by varying the amount of twist. The complex type of novelty yarn is used with two objects in view, one is to combine different fibres *e.g.,* cotton and rayon may be blended with or covered by wool or silk.

This lowers the cost of production. The other objective is to produce a novelty yarn for the construction of novelty yarns, at least one or two single yarns are used, one forms the foundation yarn known as a *base* or the *core* and the other, the effect yarn, which is wound or looped round the first one. A third yarn knownas binder yarn is often used to fasten or tie the effect yarn to the foundation yarn. These types of yarns are mostly used for drapery and unpholstery fabrics.

Simple Yarn

In the construction of simple yarn, only one type of fibre is used. The manner in which the fibres are twisted will be the same throughout the length of the yarn. Yarns, called simple, ply or cable, depending upon the number of strands they contain.

Ply Yarn

Two or more than two yarns are twisted together to form a ply yarn. These yarns are called multiple strand yarn. If two single yarns are twisted together, the resulting yarn is called two-ply yarn, if three are twisted together, three-ply yarn and so on.

Single Strand Yarn

In this, so many of fibres are twisted together into a continuous length. The yarns consist of one kind of fibre and of one colour. This kind of yarn is the one usually found in most standard fabrics for clothing and household use.

Kinds of Yarn

Knot or Spot Yarn

In this also two strands or yarns are used. One is the base, the other is utilised for forming knots or spots by an additional turn round the base yarn.

Slub Yarn

This yarn is composed of soft untwisted places at regular intervals. This may be a single strand yarn or may comprise two or more strands

Loop Yarn

This yarn comprise loops at intervals round a coarse foundation.

Corkscrew Yarn

This yarn is constructed by twisting together yarns of various diameters or by varying the rate of speed or the direction of the twist.

6

Use of Robotics Textile and Decision-Making

In this chapter use of roboties textile and decision-making has been treated comprehensively. Before a decision is made concerning the purchase of any 'capital equipment, it is best to determine, as far as feasible, the direction in which an industry is headed. So, first, let us try to realise feasible developments that could take place at some time in the future in a most modern yarn manufacturing plant.

This plant would be designed and orchestrated to produce high quality sales yarn at a cost which would not only be competitive for American mills, but also for worldwide competition. In this modern yarn manufacturing plant, cotton will be received in uncut bales with classing samples attached and comprised in moisture-proof bags. These samples will have been removed during the ginning process.

Besides the bill-of-lading, each shipment will list not only the bale number, weight, and grade, but also the fibre test results, such as two and one-half per cent span length, fibers shorter than one-half inch, fibre strength, fineness and about any other characteristic that may be desired by the buyer. The weights and fibre tests will be so accurate and the shipper will be so reputable that it will not be essential to grade and sample all bales received.

Generally, only a small percentage of bales will be tested in order to verify the quality of the shipment. Each shipment will be unloaded onto a power supplier that will transport each bale past a receiving station that will automatically register weights, apply the mill identification number; and remove any samples for testing.

All samples of the bales and the shipping lists will be sent to a grading laboratory for observation and to verify the quality of the shipment soon after it has been received. After passing the receiving station, each bale will be moved by the power supplier to a loading station where it will be picked up by a robot and taken to storage. All bales will be stored randomly in racks, each bale individually suspended.

The position of all bales with regard to bale number, weight and fibre eatres will be loaded into a data bank for future determination of each bale's use in the mixtures that will be necessary to produce optimum quality and high manufacturing efficiencies. When a bale is selected for processing, it will be removed from the warehouse by a robot on a first in, first out basis. It will be transported to an opening area where ties and bagging will be removed automatically.

The bale will be placed in a staging area for blooming prior to entering it into its proper position in the opening line for mixing. Blending and cleaning lines will be resemble to those being used today; however, they will run closc to 100 per cent, of the time and stop only for emergencies. The speed of the feeders will be varied by direct current drives. There will be no more stopping and starting of the lines as is commonly being done today.

This will give for more uniform feeding of raw material at a slower production rate in order to open and clean more efficiently. Before leaving the opening and cleaning area, let us consider the modern handling of salable and reworkable waste. All waste will be gathered as is done today, with each description of waste being handled by individual vacuum tanks. Below each vacuum tank will be a container of bale-size proportions.

This container will be of adequate strength to withstand significant internal pressure. As the level of waste in each container reaches the top, the container will be picked up by a robot and transported to a press, into which the container will be positioned.

After the container reaches its proper position, the ram of the press will compress the stock and it will be 'held in a partially compressed state by retainer dogs. When the ram is withdrawn, the container will be removed by the robot and returned to its original location under the vacuum tank to receive more waste. These steps will be repeated until sufficient waste is in the container to complete a normal size bale. Then, it will be strapped, and bagged if essential, during its last trip to the press.

This procedure will permit efficient handling of all waste and highly automated bale press operations, with only one press generally being required. Chute feeds that run continuously will feed cards operating in excess of 100 Ibs./hour, producing high quality webs.

Long-term sliver variation will be so well controlled that the overall coefficient of variation of one-yard lengths will be on the order of two per cent or less. This will be completed by sensing and controlling tile weight being fed and delivered continuously. With drafting elements situated in the calender roll section, manufacturers will be able to spin lower quality yarns directly from card sliver.

The cards will be doffed automatically and at random, with specific lengths of sliver. Robots will be programmed to pick up cans from each card, place them onto a sliver truck (eight cans from eight various cards on each truck), and transport them to a staging area for drawing for traditional operation.

This will give for excellent cross-blending and it will be a simple matter for an operator to creel the cans from the truck into a drawing frame. The drawing frames will run continuously and doff automatically. As earlier outlined, one can will be picked up from each delivery for drawing to be placed on trucks which will hold the proper number of cans for the next process.

Lap winder laps will be transported to comber creels on conveyors in sections, in order that a complete comber can be creeled with minimum loss of running time. Each succeeding process will be doffed automatically and, where required, robots will be utilised to transport full creel sections throughout the remaining processes in the card room. It will be feasible to doff roving frames automatically straight up into spinning creel sections.

The spinning creel sections will be transported to staging areas, with the roving description loaded into data banks. Spinning creels will be electronically summoned immediately prior to the time they are needed, transported to the spinning frame sections designted for creel piecing. The length of each spinning frame will be determined not only by ring size and gauge, but also by multiples of the number of spindles on roving frames.

For instance, for 96 spindle roving frames, spinning creel sections will contain roving bobbins for 96 spindles and sides of spinning frames will be in multiples of 96, *e.g.*. 192, 288, 384, etc. Spinning frames will run incessantly, with each spindle being doffed and pieced up automatically and randomly. This will be completed by an automatic piecer-doffer which will be mounted on and will patrol each side of each spinning frame.

Each bobbin will run until it becomes full, at which time it will be stopped, doffed, and the end pieced up robotically. This will permit for efficiencies approaching very close to 100 per cent (99.93 per cent for a 30-second cycle time when doffing and piecing bobbins that run 12 hours). With the low cycle times that will be developed by the random doffing and piecing of spindles, spinning frames will be designed with more narrow gauge and ring sizes very less than two inches for medium count yarns; this will permit much higher front roll speeds than those normally being run today.

The doffing and piecing robot will deposit bobbins at the end of the frame into doff boxes which, when either full or as needed, will be picked up by transport robots and taken to the winding area where the yarn will be staged.

The description of each spinning doff will be loaded into a data bank and, when required at winding, the spinning doff once again will be picked up by a robot on a first in, first out basis and transported to a winding loading station. It then will be positioned into a hoist, the doff box will be elevated, and then will become a hopper, feeding yarn to each winder spindle automatically. The yarn will be cleared while it is being wound, similar to today's practice.

However, we will receive much more fruitful information from the clearing equipment, which will be able to designate various classes of defects either per unit length or per unit weight. In effect, the electronic clearing system will become an effective laboratory tool capable of monitoring virtually all yarn defects on a 100 per cent basis. Most yarn will be wound onto parallel tubes rather than tapered cones because of new developments in creels and storage feeds that do not need tapered packages.

This will permit greater machine efficiency both in winding and in knitting, because of the larger packages that can be produced with fewer winding defects than by today's standard for tapered packages. After winding, all cones will be doffed automatically and transported to an area where they will be packaged by robots, with all cases being strapped, weighed and labelled automatically.

All cases will be stored randomly in warehouse areas, with the description of yarn in each carton being loaded into a data bank before to shipping. When shipping instructions are received, computers will select cartons of each yarn description to be shipped on a first in, first out basis and, through the use of robots, they will be transported to a loading area. In the loading area, whole shipments will be made up prior to the truck's arrival.

Each shipment will be encased in portable containers mounted on rail tracks which lead directly to the door of the shipping dock. When the carrier arrives, ridged connections will be made with the transport track to a track which will be into the bed of each truck. After the truck is positioned at the loading dock, the whole shipment will be loaded and ready for departure within five minutes. This will give for much more efficient utilisation of manpower and equipment, both in the manufacturing plant and for the carrier.

Computers feeding on-line printers will make up bills-of-lading and invoices which, will the utilisation of communication modems, will be loaded in computer files in the accounting areas of the manufacturer, the carrier, and the customer. No invoices or checks will be mailed. Payments will be made electronically by transferring balances from the customer's bank account to the manufacturer's bank account, with updates on all balances, payments, and receipts being made electronically in each office each day.

Now what we have described may appear far-fetched; however, in most instances, the technological art at each point of this imaginal operation is either being practiced today or is at some stage of development. As you have already recognized, much of the operational and most of the service work done at each process may be handled by robots.

The complement of people needed to run this plant will be highly skilled technicians and observer/operators whose primary liabilities will be inspection and correction of malfunctions that may occur. This is where we come to the most vital and potent part of this operation-people. Without properly motivated, trained and skilled people, no operation can be a success.

Highly trained employees, skilled in electronics, microprocessors, computers, and developed electronic systems will be-needed to keep this plan operating smoothly and efficiently. All levels of management, in addition to being thoroughly skilled in manufacturing technology and modern management procedures, must also be needed either fo be skilled or at least thoroughly familiar with electronic and computer systems.

Indeed this represents a challenge to all who hope to be successful in the future. With a plant so highly automated, we must not overlook the importance of human requirements. It is imperative that we percept these individuals must not only be highly skilled and self-motivated, but they also must be developed through a quality of life that transcends all fragile relations between man and man, and man and machine. How soon may we hope to be able to build and successfully operate a plant similar to that just described? It may be sooner than we think.

Hopefully, we are in a financial position and are operating successfully in an area which allows us install capital equipment as we know that it will be essentail in the reasonable future, either because of quality requirements, improving the working atmosphere of employees, cost reduction, or for environmental consideration.

Of course, we must not allow ourselves to be lured into a position whereby, with the subsidence of direct operators and service employees, we end up with a host of engineers to keep our plant running. Manufacturers of robots and likewise equipment must realize that for their product to operate successfully, all parts, which include both the electronic and mechanical actions, must be thoroughly and very reliable enough to operate on a continuing basis–24 hours each day.

Excessive down time on equipment of this type cannot be tolerated. Faults which occur, although they may be caused by high tech difficulties, must be readily repairable by employees who are not so technically qualified. Operations in some areas of manufacturing as described are possible today and each of us with open minds certainly should give full consideration to going ahead and "testing the waters", so to speak, not only to improve quality, productivity, and to subside cost, but also to start developing our people and our minds for future events in our business which may be very resemble to what has been described.

The use of robots in textile manufacturing will probably develop in two areas. One will be for those jobs which are laborious and repetitive; the other will be in those operations which make treat improvements in quality and very substantial increases in productivity. Laborious and repetitive jobs which can be handled by robots in the future are those such as receiving and storing cotton, opening bales prior to blending, transport of stock throughout the plant, and packaging.

Operating robots which improve quality and productivity will be resemble to those used today in automatic splicing on winders and automatic piecing on spinning frames. Robots are being used in textiles today and substantial developments are already being made. Since we operate in what may commonly be described as free enterprise economy, the future use of this new technology will develop as the economics of each area of manufacturing dictate. Undoubted about it, all of us can look forward to exciting developments. Challenges, rather than being reduced, are greater and more enhanced; and definitely, those who succeed will be rewarded by the great future which can be expected for our industry.

The art of textile manufacturing has advanced from the starting of civilization. It is not as to nishing that this should have been the case, as life itself was linked to the utility of textiles and it behaved man, at whatever time, to do the best that he could to procure textile materials as he gave for himself the essentials of life and welfare.

ASPECTS OF ROBOTIC MATERIALS HANDLING IN TEXTILES

By the time of the Bronze Age, looms had been invented that used many rigid materials, such as wood and horn, to position warp yarns in a more or less stationary configuration so that one could more efficiently insert the filling amongst the warps to produce the woven textile. Whorls and spindles had been invented to help in the forming of yarn and its packaging. The over-all processes of textile manufacturing had become sufficiently institutionalized to where textile manufacture was a treat fireside routine engaged in by members of the family, mainly the females.

These textile-making processes included the transfer of the crude fibre into yarns, initially with the aid of the whorl at the spinning wheel. The yarns then were placed into the hand loom as warps and later onto the bobbins that were loaded in the shuttle, which was thrown backwards and forwards through the yarns of the warp shed to insert the filling into the fabric.

The primary power related to this operation involved the human effort of beating up the yarn thus inserted into the fabric which, in turn, was wound up onto the takeoff device. As the art of weaving progressed, it soon became apparent that more yarns were required to supply a loom installation than could be made by a single family.

A loom at the hearthside needed a large number of daughters constantly engaged in the production of yarn to keep the fabric production from the loom at the level needed by the family needs. This state of affairs obtained largely up to the inception of the Industrial Revolution.

There was, however as tonishing time when a premature motion towards a factory workplace occurred in ancient Rome. It was related to the production of woollen goods mainly for togas, wherein women from various households met collectively in a workplace to produce yarn and woollen fabric for clothing of the citizens of Rome.

This movement was interrupted with the fall of the Empire and it needed several hundreds of years to elapse before this early movement towards the factory, was to become the generally accepted way of life. The reintroduction, of the workplace at the end of the Middle Ages sparked the renewal of the movement towards the modern industrial society.

As the introduction of a factory workplace was renewed by the Industrial Revolution, the factory quickly became related to a region with easy access, to power to run axles, which in turn gave some of the energy for yarn twisting and for loom operation especially filling insertion and the beating up of the filling as it was inserted into the warp sheet.

The factory provided the requisite space allocation for yarn production so that a balanced production could be achieved. In France and other parts of Western Europe, spinning halls were introduced for the specific making of yarn, in which young girls were utilised as labourers. They kept production rates uniform among a large number of spinning positions by timing their wheels as they sang multipart rounds, such as "Pop Goes the Weasel."

As this extra space was given, so too were mechanisms for the handling of the yarn and conveying it from a point of manufacture to a point of use, that is, the point of insertion into the fabric.

While much human labour was still utilized in connection with the early factories of the textile industry, the development of factories took place in concert with the introduction of nonhuman handling devices wherever technological ingenuity was sufficient.

It is this feature of the early textile workplace that contained the germ of automation and robotics as we know the processes today. By the time that Jacquard looms had developed to the point where one could produce highly-patterned woven tapestries utilising machinery, the textile manufacturing process had been integrated with what were, in crux, early computers; these machines utilised punched cards to give the complex motions that were imposed upon the loom heddles.

With this programmed control of weaving machinery, figured tapestries and brocades were obtained which, in few cases, closely approximated the finer handwrought works of art that had been produced mainly in the nunneries of Western Europe. With the expansion of manufacturing processes immediately preceding World War II and throughout that period, techniques were perfected for using labour together with advanced machines in the creation of rudimentary "self-thinking" processes.

It was a fairly simple matter to devise lathes and grinding machines with human operators for the production of cylindrical shaped objects, live projectiles and machine screw threads. A simple extension of this process rapidly evolved in which the product was formed automatically and then transferred to an inspection station that confirmed the precision of the production. On the ground of a "go/no-go" measurement, the product was either accepted within a given channel or, failing the "go" measurement, transferred into a reject channel.

It was a simple matter at that point to automatically transfer the product back to the manufacturing site and subject the item to reshaping if that were feasible. Those items which were produced outside appropriate tolerance levels were, of course, rejected permanently.

It was not a difficult matter to extend the go/no-go production inspection scheme to the point where the role of the human operator was progressively minimized to that of an overseer, who made sure that the automatic setting and conveying steps in the processes were operating properly in their appropriate sequence.

From this kind of process development, it was a relatively short time until there was developed not merely a go/no-go inspection surveillance of'the production, but instruments that could be utilised to measure chemical and/or mechanical disturbances outside of an accepted tolerance. The measurement was utilised to activate appropriate mechanisms to move the process back within accepted tolerances. The role of the human was thus removed from the production process, and what several have termed the second industrial revolution was clearly recognized.

The second industrial revolution produced devices which replaced not only the muscle of man, as took place at the starting of the first industrial revolution but also replaced man's mental capacity and ability to follow with appropriate non-human mechanisms, so that now the production essentially was completely automatic. A "generalized robot" had been built to replace man's immediate role in the processes of manufacture, In the textile industry, one of the first clearcut examples of this total replacement involved the recognition of off-specification product quality.

The process was equipped with a means for "feeding-back" that recognition to a station which adjusted the controls so as to obtain the needed specification. This early textile process control involving feedback is found in the control of temperature and of size bath concentration, which are used to coat the warp yarns to be used in the loom. This kind of process control with its inherent feedback and essentially labour-free character was clearly in place by the end of the 1950s.

By this time the, second industrial revolution was gathering momentum throughout America. In the textile sector again, semi-automatic production control, developed in small steps in connection with production of synthetic yarns, gives a second example of production with feedback control. This took place in connection with the control of the viscose dope in the production of viscose filaments. In the art of filament handling, there are a number of processes that comprise the maneuvering of filaments after processing to further enhance the yarn-like character of the filament product.

This involves the processes of texturizing which, in general, again involve not only automatic handling, but also a programmable sequence of temperatures that is imposed on the yarn as it is maneuvered through the processing equipment in order to achieve the desired snarled, interlaced, character. Such machines are elaborate robot factories.

They not merely handle material and call for various thermal treatments of it, but they can be programmed to produce a great variety of processes, thus giving the textile manufacturer an enormous inventory of yarn-like items made completely from the synthetic filament technology of the 20th century. Also in the case of filament yarns, a bundle of filaments appropriately interlaced by means of air jets produces a structure that can be subjected to twisting, plying or air texturizing to produce a looped and whorled structure that can be plied, plaited and/or twisted further.

This yields a continuous filament structure that has surface characteristics resemble to those of the conventionally spun yarn produced from staple fibre. Along with these developments of filament technology, newer types of spinning processes have evolved, comprising conventional staple products.

The open-end spinning process involves the introduction of staple fibre by means of an air stream into a yarn forming chamber that provides intermingling between the filaments and a twisting of the product to produce a yarn which is similar to more conventionally spun yarns.

The primary mechanisms in these machines with their robotic characteristics are due to European manufacturers of textile machinery. The beginning of these processes comprises suction cups or mechanical fingers that can be moved over the surfaces of the fibre bale to progressively "eat" the material off the bale and convey it through air handling passages to the card.

The product of the carding and drawing prodesses is fed into the open-end spinning machinery. All of these instances give rise to a certain perspective concerning the current art in the handling of textile materials.

It has attempted to give not only a retrospective review of many aspects of the production of textiles, but also to indicate that, as the history of textile production has unfolded, the introduction of advanced machinery occurred. Ultimately that machinery was such that it performed in several aspects as if it were "thinking", so as to produce a product without the advantage of essential human labour.

This is a somewhat more complex view of the robot than that discussed. Accordingly, one is forced to consider that a highly articulated mechanical device with appropriate sensing elements to permit it to "see" as well as to "think" perhaps is only a simplification of the more modern production techniques which will develop in the future.

In the textile industry it is already feasible to see places where the robot and the production mechanism may be thoroughly integrated. Regarding certain modern plants the conventional spinning frame is encircled by a travelling device which has articulated motion and is capable of sensing whether or not a given yarn is broken or a bobbin has been filled.

At these times the device plucks the filled bobbin and places it on a conveying belt to the production output station or ties up the ends which have broken as a consequence of a failure in the spinning process. Alternatively, mechanical devices for conveying yarn packages to looms can be seen in the automatic guided vehicle inevitably small cars moving on a track operated electronically by computers. These are activated when the loom senses that it is about to run out of yarn supply, and an inventory system is tapped automatically.

The necessary yarn is placed upon the car, which then is sent in response to the request of the loom station to that loom where the yarn is required to sustain continued production. A variation of this process, using much of the same equipment, is activated when the loom has produced a full beam of fabric, at which time the loom is doffed and the finished fabric rolled onto the conveying trolley and transported to the production shipping point. This process can be realised as integrated into a finishing or other later processing plant.

All of this more highly integrated production turns the total plant into a complicated robot which consumes raw material at its receiving dock in the form of fibre or filament and manipulates it through mechanical conveyance and sensing stations, the results of which are fed into computers which iii turn activate extra conveying and moving devices to manipulate the fibre into yarn and ultimately into fabric. The prime human content of all of this is that of the overseer of the process.

From the economic point of view, this new-labour factor (that of the robot factory) has a fixed cost with 'a particular maintenance requirement, and the new human labour need involved in the production is now that of a more complex reasoner or decision maker who monitors the overall activity of the production unit.

The opportunity for this decision maker to engage in entrepreneurial activity is comparatively limited, but there still is an essential and major difference in this human labour input, compared with that of the labour that operated factories at the beginning of the Industrial Revolution.

The second industrial revolution has, in fact, separated that which is machine-like from that which is intellectually more stimulating from human labour content; it has permitted the human component to perform in a somewhat more elevated environment and transferred the drudgery of the process to that of a nonhuman robot slave. The techniques for the production of material things underwent a substantial change at the starting of the first Industrial Revolution.

The first revolution can be thought of as the introduction of automatic transfer of power from process unit to process unit conveying work material from machine to machine, or performing automatically certain work tasks utilizing a power source which is available in the workplace.

The second revolution culminates in the recognition of automatic devices which are programmable to stimulate through feedback loops several aspects of this power transmission to complete process modifications upstream of the sensing measurement.

In this sense the two industrial revolutions can be thought of as end-points of a scale of production which has been incessantly varying throughout history from the first towards the second. The manner in which human labour is used has changed substantially. In the starting it needed the collection of workers from their homes into the factory workplace and the organization of these workers within the scheme of production.

The second revolution leads to a displacement of several of the worker positions in the production process but introduces the need for more highly skilled service employees to keep the process operating at the desired level. In the earliest phases of robot introduction into textile processes much of the traditional yarn making and weaving process details are maintained in the newer technological scheme, and the automatic devices with their feedback loops are simply embroidered around the existing processes, thereby reducing the number of human labourers that earlier were employed in the chain of production.

Devices for the automatic retying of yarn ends on spinning frames where breaks have occurred at particular spindles represents one of the simplest instances of the substitution of automatic equipment for human labour. In recent past, robot-like devices with limited articulation have been mounted on carriages which move around spinning frames and offer not only automatic relying of broken ends, but also give doffing and the starting of spinning by the throwing on of yarn onto an empty take up package.

In this way the production of yarn can be made entirely automatic from sliver to the wound-up package. It in turn can be conveyed automatically to a next stage in the production chain. Extra robots are under development that will permit the packaged yarn to be picked up from a supply source and conveyed from the point of yarn manufacture to the point of use within a weaving or knitting process or the like. The robot transfers the yarn, puts it into the next stage of the textile production chain and starts the production process. These motions need substantial amounts of movement through space and a series of sensing devices that control the subsequent motions. Another textile-related aspect of the second industrial revolution is found in the elaborate automation involved in the maintenance of material inventories. As an example, a picking device can traverse a predetermined set of fibre bales to produce the essential tufts and convey them aerodynamically to the card for the formation of sliver which is then fed into spinning devices, for instance, open-end spinning processes.

The automatic feeding of these devices can be structured in such a way that one can sample statistically the contents of a series of bales so as to achieve the eventual in blending from a given supply of baled staple fibre. Similarly in the case of continuous filament yams it is feasible to receive at the input of a factory site the large cheeses or straight wound packages, which are transferred to a series of trolleys.

When an appropriate encoding is provided at the loading station these convey the yarn to an appropriate spot within the inventory area, to be withdrawn on demand. Such production units are now available wherein hundreds of looms can be found on a given weaving shed floor with relatively small amounts of human labour, on the order of ten workers per 300 looms.

The looms are regularly supplied on demand by trolleys working from an inventory containing the necessary yarn components for the fabrics being woven within the factory at a given time. The human labour content is primarily that of correcting various weaving errors which originate from yam irregularities and giving machine maintenance.

In some cases this minimal labour force must sense the approaching need of a given machine for input supplies and signal that need to the inventory delivery system. This type of automatic production of necessity needs either the automatic sensing that a beam has been filled with woven fabric or that the pirn supply for the feeding of the shuttles is running low and needs resupply. All of the processes briefly alluded to above represent textile processes that have developed through the first and second industrial revolutions and thereby contain a minimum of human labour. Further, labour which is inherent to the processes tends to be of a rather very trained and sophisticated variety.

Individuals must understand the nature of the automatic sensing devices and the feedback loops, and the controls which are passed through such loops, as the processes for self regulation continue to operate throughout the production cycle. Further the complex mechanical and electrical mechanisms which are included in this production system comprise both elementary mechanisms and highly sophisticated electronic and electromechanical devices.

One can quickly infer from this that the processes for programming these processes are indeed complex and comprise a complicated set of inputs which can and are changed from time to time as the production is tuned towards higher and higher quality levels or towards greater and greater productivity with some acceptable fault level.

Automation Activity

Due to the attention being provided to Japanese automation activity, an impression exists that United States sewn products and equipment industries are not active in trying to improve productivity through automation. This is not true.

Since the 1960s there have been a variety of attempts to attack both particular and common joining problems. Joining limp fabric semi-automatically or automatically presents a variety of problems which have been and are a constant challenge to equipment manufacturers and the industry as a whole. It is almost not possible to detail the wide variety of efforts which have been made in this area.

Basically it appears that the Japanese have proceeded to collect patent information worldwide, organize it and make an orderly assault on creating a sewing system. In making this assault, teams of government technicians, university staffs and industry engineers roamed the world gathering main sewing system information. A high proportion of what they are studying is of U.S. origin. Why then haven't the American engineers mechanics innovators succeeded in developing a sewing system? What have they been doing all this time? What do they require to achieve success?

New Techniques of Sewing

Assembling apparel, shoes, tents, curtains, furniture and auto seat covers are but a few of the several activities in which thread or thread substitutes are utilised to join seams. Joining parachute. seams represents a substantial difference in handling techniques than joining a styled outerwear jacket - even though both products may use the same material and kind of joining equipment.

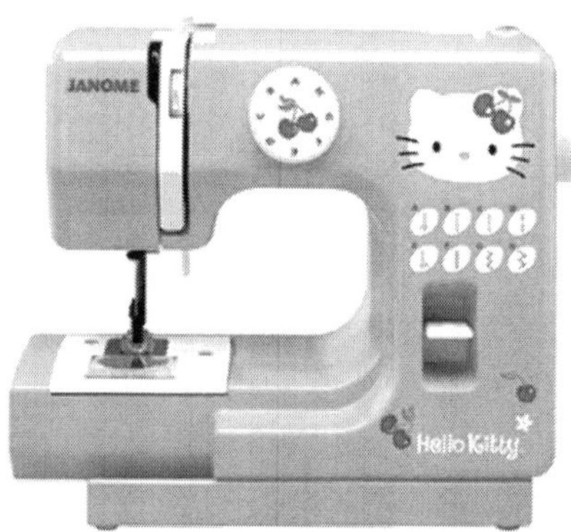

Fig. New Techniques of Sewing

As joining activities are investigated, the response to solving problems tends to become product oriented. That is, the problems of assembling a product are fragmented into a series of digestable steps. Ordinarily, production people and industrial engineers solve assembly problems by copying the way they have been earlier solved. Equipment firms provide a quick reference point for establishing sewing production lines. When production lines are set up under time restraints, the opportunity for creative innovation is limited. The sewn products industries have consistently sought to provide better equipment to satisfy both individual and broad industry needs. Modifications of equipment generated by the mechanic/innovator to solve readily problems are often incorporated into later versions of that equipment by the equipment manufacturer.

Handsewing, either actual or with machines, which requires highly skilled operators may be perceived as either a problem or an opportunity to seek solutions. Solutions have tended to create particular equipment, sometimes of limited general usefulness.

Adding of Feeds

The Federal Stitch Catalog lists a succession of alternative stitch formations in common use. Machine suppliers compound this complexity by adding a variety of feed and material handling mechanisms. Needle suppliers give a wide range of alternate shapes and sizes for use on different products and materials.

Accessory suppliers display a wide range of folders and button handlers to enhance material handling and joining quality. Joining is a highly complex activity which deals not only with several equipment alternatives but also with problems of seam strength, decorative stitching, and in some cases, the creation of air tight seams.

When the complexities of sewn joining are considered in a world of changing fibers, fabrics, wovens, non-wovens, knit and molded goods, it is not surprising that fully automated factories have yet to develop in the sewn products industries. But, this does not imply that some steps towards a higher level of automation have not occurred. The problem is being studied.

Handling

Probably the most frustrating facet of joining parts together evolves trom the wide variety of materials used in creating sewn products. A substantial portion of existing equipment activity is devoted to developing or improving existing material handling devices.

Almost every study proves that moving and positioning materials to, under, and away from the joining process is the chief labour time-consumer within the sewn industries. The Japanese Apparel Automation Project selected thirty material pick-up devices for study in their “System J” project. Twenty-one of these devices were patented in the USA, five in West Germany, two in Great Britain and two in Japan. From this study, a new technique comprising of a modified pick-up device and “stiffening of material” was developed.

What is significant in this study, insofar as the American apparel industry is concerned, is that better than two-thirds of all the inventions considered in this report were developed in the United States. The Japanese preliminary study represents a relatively small portion of all the handling devices which might be used for bringing together and positioning parts to be sewn, bonded or otherwise joined. Omitted are overhead and belt conveyors, sensors and switches, indexing equipment and several other devices, in full or pertial use within the sewn products industry.

Matching

Matching one part to another in the joining process is being approached by the machine industry in a variety of ways. Most of the methods use an operator to feed a clamp or clamping point which in turn actuates a series of positioning and joining actions.

These semi-automated activities are controlled, cam and electronic controlled, or electronic controlled. Because the methods use equipment to actuate the joining operations, an operator can generally handle more than one machine. Skill levels are reduced while productivity is increased.

In some cases a series of activities are combined in one sewing centre. Pieces of equipment are combined to do more than one sewing operation or to accomplish intricate jobs semi-automatically.

Methods of Accomplising

As properties of materials change, the methods of accomplishing product design features also change, While there is nothing new in shaping materials through steaming, pressing, stretching or moulding the characteristics of several materials allow for a broader use of these techniques today. Moulding of products ranges from bra cups through wet suits to automobile seat covers.

Several of the moulding or shaping techniques include adhesive supporting materials to improve joining bonds and to improve retention of shape. When these techniques are considered as significant displacement of other sewing activities, another range of complexity faces the designer/production/engineering activity.

Shaping or forming products instead of sewing them has seen broader use over the past 15 years, and is achieving increasing acceptance. Items like slacks, uniforms and dresses have been formed in whole or in part.

Patents have been granted in the United States, Great Britain, West Germany, Japan and Czechoslovakia for making semi-complete and complete garments. Techniques as varied as spot welding by heating a bow on lingerie through utilising sound to cut buttonholes are starting to achieve acceptance. Mixture of natural and manmade fibers offer the innovative designer opportunities to create new looks and better products through the use of moulding and bonding techniques.

Limited Production

Most sewing products firms are small and tend to produce limited volumes of goods for splintered markets. Tiny firms with limited management are not likely to make investments in equipment research and development. The fragmentation of this industry is, in itselt, a barrier to joint equipment manufacturer luser efforts.

The sewn products industries represent a low investment sector of the economy. seldom no investment is made in equipment; the manufacturer rents it.

In this environment of small, diverse users catering to limited markets, high speed and high volume equipment has a doubtful market. Sewn products research and development activity is largely conducted by the chief sewn products firms.

This R&D generally takes one or more of four forms:

- In-house R and D,
- In cooperation with an equipment firm,
- Employment of a research firm capable of producing a prototype
- Employing a university.

Two other forms of R and D investigative activity have been sponsored by apparel and equipment firms:

- The Technical Advisory Committee (TAC) which comprises members of American Apparel Manufacturers Association (AAMA)
- The now defunct Apparel Research Foundation (ARF) which consisted of 375 dues-paying members and which received some R and D money from the Government.

Apparel Research and Development R and D

Apparel R and D tends to be cyclical and specific. Departments are created and funded when business is good. Projects are generally highly product oriented, usually devoted to either improving quality, reducing cost or both.

A high proportion of these projects are basically short term in nature. This kind of R and D can often lead to a competitive edge in the market place. Generally, the risk of investing in this kind of R and D is relatively low since targets tend to be specific, limited in scope and well planned. In the earlier example of separating, picking up and positioning materials, 21 of the patents reviewed by the Japanese were U.S. patents.

One patent was held by a research firm, one by an individual, and the rest by equipment firms. Review of patent records in other areas indicates a similar pattern of activity. It should be kept in mind that several large firms, such as Farah and Haggar, do not usually seek patents yet pursue a reasonably active R and D programme.

Besides, several smaller firms whose mechanic/innovator develops a device pass the item to an equipment firm to receive a production item. This increases the equipment firm's patent volume and reduces the activity attributable to the sewn products industries.

Used Equipment Firm Cooperation Firm

This form of R and D again is generally specifically directed at an opportunity. The user is a major prospect for the equipment firm, but ordinarily the equipment firm also perceives a market beyond the user.

Projects evolving from user/equipment firm cooperation provide for user testing and user first call on the equipment. Success in these joint ventures has been spotty, with most of it in the area of combining jobs. Probably the best known joint venture is that of Union Special and Cluett/ Pfaff in developing semi-automatic assembly of shirts. This venture covered a span of several years and resulted in several innovative machines which are now on the market.

There were some ideas which did not work out at that time, primarily because the electronics were not available. In recent year, Kellwood/Union Special, Lee/Jet Sew, and Haynes/Union Special provide a sample of firms working together on specific problems ranging from automatic bottom hemming to combining three jobs into one. Both development and failures resulted. However, most of the projects obtain most of the objectives.

Investigation of Equipment Objective

A number of research firms are in the position to investigate an equipment objective, determine its possibility and produce a prototype. Generally than not, this type of activity has, in one way or another, been jointly financed by several user firms. This approach to R and D was popular during the late 1960s and early 1970s. As an instance, Blue Bell, HD. Lee and Levi Strauss jointly worked with Battelle Institute on equipment to full-fell jean seams semi-automatically.

Some positioning and pulling mechanisms were developed which evolved into work aids, but the chief programme of producing a smooth full-felled seam semi-automatically was never solved. The 375 members of ARF together with the Government commissioned Arthur D. Little Co. to investigate limp material, handling devices. This project was successful through the prototype level.

The patents were part of the public domain. Unfortunately, equipment producers who move forward with this system to a production model do not have any protection from competitors. Therefore, the results hoped for were not obtained. Automation Equipment Labs (AEL) approached the sewn products industry seeking backing for a series of automation projects. The firm received rather a bit of backing. The expense related to attacking a broad spectrum of joining problems developed into more than the backers felt they could afford. As a result, they dropped out of the semi-joint venture. These instances represent just a few of the joint ventures in which users have attempted to move forward. They do have several things in common:

- There is a limit to how much they can afford to spend on R and D and still stay in business.

- The sewn products industries identify a requirement for better ways of doing things.
- They have been interested to devote time, manpower and substantial resources to projects.

Universities

Sewn products firms have supported university research on a limited basis. Majority of the industry's support has been directed towards providing equipment and funds supporting courses in apparel activities. Sometimes, advice is sought on chemical or mechanical problems, usually in conjunction with a specific in-house project. Support of university research tends to derive from individuals within the industry rather than as funded corporate projects.

Role of TAC Committee

Each year the TAC Committee reviews the state of the art in some area of the industry. Reports are prepared and provided to AAMA members. Basically, the reports are a summation of what the committee has found to be in existence within a specified area.

It gives the industry an opportunity to bring itself up-to-date on either equipment or operating systems according to the subject selected for the year. The reports also generally indicate trends and give management with useful background information.

Function of ARF

The function of ARF was resemble to that of TAC in some respects: its operation, the foundation collected information on equipment and trends. A quarterly report was issued to its members. ARF also kept track of government activity which might be of interest to the industry. It obtained funds for some research activity from the government.

Among the last proposals it made to the government was one to the President's Domestic Council on October 30, 1971. This proposal reviewed the state of the apparel industry, the activity of the Japanese apparel industry and proposed a series of counter measures. Among the proposals was a request. The 375 members of ARF together with the Government commissioned Arthur D. Little Co. to investigate limp material handling devices. This project was successful through the prototype level.

The patents were part of the public domain. Unfortunately, equipment producers who move forward with this system to a production model do not have any protection from competitors.

Therefore, the otucomes hoped for were not obtained. Automation Equipment Labs (AEL) approached the sewn products industry seeking backing for a series of automation projects.

The Firm Received Rather a Bit of Backing.

The expense associated rather attacking a broad spectrum of joining problems developed into more than the backers felt they could afford. As a result, they dropped out of the semi-joint venture. These examples represent just a few of the joint ventures in which users have efforted to move forward. They do have many things in common:

- They have been interested to devote time, manpower and substantial resources to projects.
- The sewn products industries recognize a requirement for better ways of doing things.
- There is a limit to how much they can afford to spend on R and D and still stay in business.

Support to University Research

Sewn products firms have supported university research on a limited basis. Majority of the industry's support has been directed towards giving equipment and funds supporting courses in apparel activities. Occasionally, advice is sought on chemical or mechanical problems, generally in conjunction with a specific in-house project.

Support of university research tends to derive from individuals within the industry rather than as funded corporate projects.

Technical Advisory Committee

Function of TAC

Every year the TAC Committee reviews the state of the art in some area of the industry. Reports are prepared and provided - to AAMA members. Basically, the reports are a summation of what the committee has found to be in existence within a certain area.

It provides the industry an opportunity to bring itself up-to-date on either equipment or operating systems according to the subject selected for the year. The reports also generally indicate trends and give management with useful background information.

Apparel Research Foundation (ARF)

The function of ARF was similar to that of TAC in some respects. During its operation, the foundation gathered information on equipment and trends. A quarterly report was issued to its members. ARF also kept track of government activity which might be of interest to the industry.

It obtained funds for some research activity from the government. Among the last proposals it made to the government was one to the President's Domestic Council on October 30, 1971.

This proposal reviewed the state of the apparel industry, the activity of the Japanese apparel industry and proposed a series of counter measures.

Among the proposals was a request for funds to develop a preliminary pilot plant design.

The proposal made suggestion that the plant be designated to make military clothing utilising experimental techniques. The proposal was not accepted.

7

Vegetable Fibres

COTTON

The origin of cotton can be traced back to India the original home of the best and finest cotton produced over world. The cultivation of cotton gradually spread throughout Asia to few parts of Africa and eventually, to the Southern States of U.S.A., which now grow more cotton than the rest of the world. Cotton is the white, downy fibrous substance covering the seed of the cotton plants.

Fig. Cotton Field

The seeds with this covering are encased in pods which grow on the cotton plant and burst open when ripe, disclosing the white, downy covering of the seed now grown into cotton fibres. The cotton plant grown in the tropics requires a climate with six months of summer weather, to blossom and produce the pods. The cotton fibre is the shortest of all the textile fibres. Its length varies from 8/10 of an inch to 2 inches.

Cotton with short length fibres is technically called 'short staple' and the one with long fibres is called 'long staple'. The latter fibre is valued more as it is utilised for making the fine qualities of cloth for which it is specially suitable, as it is easy to spin and produces a strong, smooth yarn. It is also suitable for mereerisation.

PROCESS OF COLLECTING OF COTTON PODS

Fig. Collecting of Cotton Pods

When the pods burst they are picked from the plants and collected. If left on the plant, the cotton gets discoloured and dirty by explore to the plant, the cotton gets discoloured and dirty by exposure to the sun and weather.

Process of Ginning

The process by which the seeds obtained from the pods are separated from the cotton fibres covering them. This process is carried out by means of machinery, and the seperated fibres have now become 'Cotton'.

Practice of Baling

The cotton is then pressed into bales which are wrapped with jute sacking and bound with steel bands. The cotton is then supplied to the mills in bales.

Cotton Preparation before Spinning

In the mills, the bales are opened and the cotton is pulled out and beaten to remove loose dirt or any other foreign matter loosely present in it. The opened out cotton is then compressed into a sheet called lap.

Process of Carding

The lap on a cotton sheet is passed through a machine called a 'card'. In such machine, the cotton is thoroughly cleaned of ail attached dirt and foreign matter and the matted fibres are separated and laid nearly parallel to each other.

Process of Combing

This is an improved combination of the carding process. In this, all short fibres are elirr inated and even long fibres are laid more parallel, thus forming a film like sheet of fibres.

Fibres which are carded and combed are of more even long and smoother than those which are only carded.

Process of Slivering

At the end of the 'carding and combing' processes, the film-like sheet is drawn into a strand about one inch in diameter called the 'Silver', which is collected through a coiler into a can. At this stage, the cotton is ready for spinning.

Special Features of Cotton

Its formations

The cotton fibre is composed chiefly of cellulose which constitutes 88-90 per cent, water 5 to 8 per cent and other natural impurities.

Construction

When raw, the fibre has a tube-like structure containing sap. When the fibre ripens, the sap dries up, the tube collapses and the fibre becomes like a flat, twisted ribbon. A central canal, the lumen runs through the fibre and in rich cotton it appears like an irregular loop. In dead cotton it is practically absent. Such fibres are weak and brittle. Cotton fibre, therefore, has no lustre or elasticity.

Fibre Shrinking

The fibre itself does not shrink but the fabrics: made with it which have been stretched in the finishing processes do.

Impact of Moisture and Friction Cotton

It is not affected by moisture or friction and is stronger when moisture content increases. Therefore, it lends itself to the washing process.

Hygroscopic Moisture

Cotton does not hold moisture so well as wool or silk, but absorbs it and so feels damp, much more quickly. It also rapidly spreads over throughout the material.

Good Conductor Heat

Cotton is a better conductor of heat than wool or silk but not as good as rayon.

Dyes of Cotton

Cotton takes in dyes better than linen but not as readily as silk or wool. If a mordant is utilised, cotton is easy enough to dye. Mordant colours, direct or substantive dyes should be applied to cotton.

Acids and Alkalies Action

Strong acids will destroy the fibres soon. Dilute inorganic acids will weaken the fibres and if left dry will rot it. Thus, after treatment with acidic solutions, cotton articles should be thoroughly rinsed. However they are very little affected by organic acids.

They are also very resistant to alkalies, even strong caustic alkalies at high temperature and pressure.

In 18 per cent NaOH cotton fibres swells, spirals, twists uncoil and shrinks and becomes thicker. The resultant fibre is smoother, lustrous, stronger and has increased water and dye absorption.

Bleaching Impact on Cotton

All bleaches can be safely used on white cotton. Linen is believed to have been utilised nearly 10,000 years ago by the European Neolithic people. Fragments of the cloth have been discovered in parts of Switzerland, the home of the Neo-lithic people.

In Egypt, the Mummy cloth or the material wrapped round the preserved dead (mummies) thousands of ears ago has been findout to be linen.

Thus, linen has been an important fabric in the past. Even today, a modern house-wife is proud of having lipen for use in her household.

Linen is the bast fibre obtained from the inside of the stems of the flax plant.

Flax is grown inmoist, temperate climates. Most of the European countries and Egypt cultivate it.

Its Requirement

When the stems of the plant turn yellow at the base, and when the seeds turn from green to pale brown, the plants are pulled out by the roots. Pulling out by the roots gives long, unbroken fibres.

Process of Drying and Rippling

The flax after it has been pulled out is tied up in bundks and left to dry for a few days. The leaves and seeds are then removed from the stems by a process called rippling.

This is done by passing the head of the plant through a coarse comb or a machin. Care is taken not to break or injure the stem. After the removal of the leaves and seeds the stems are again tied up in bundles.

Some Other Small Fibres

Hemp Fibres

This is also a bast fibre stronger than linen, or jute, and is dark brown in colour barring the Manila hemp, which is white. It is not used. For weaving fine cloth but is used for making gauzes, ropes and webbing.

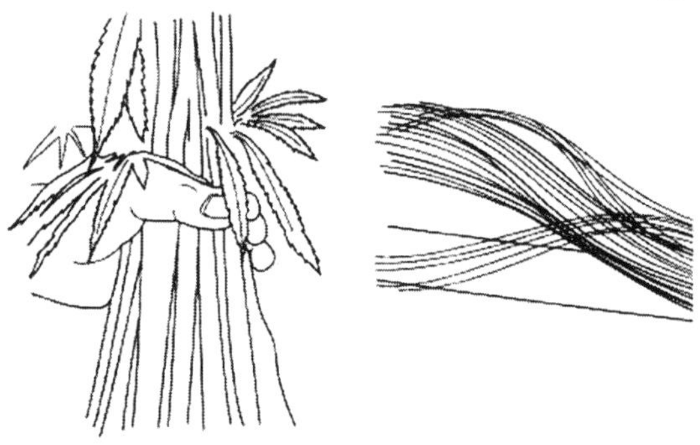

Fig. Hemp Fibres

Jute Fibres

Fig. Jute Fibres

This is also a bast fibre of plant 'Cochorous Capsularis' that grows in India, to a height of 12 ft. The fibre is prepared in the same way as flax fibre.

These fibres are weaker than linen, are short, lustrous and smooth. Jute is affected by chemical bleaches, and so it is never made pure white. It is utilised for floor covering, gunny bags, binding threads, and also used for dress materials and saris.

Kapok

Fibres is a cotton like fibre. It is get from a tree grown chiefly in Java, West Indies, Central America and India. The fibre is finer than cotton, is silky in look but is not suitable for spinning.

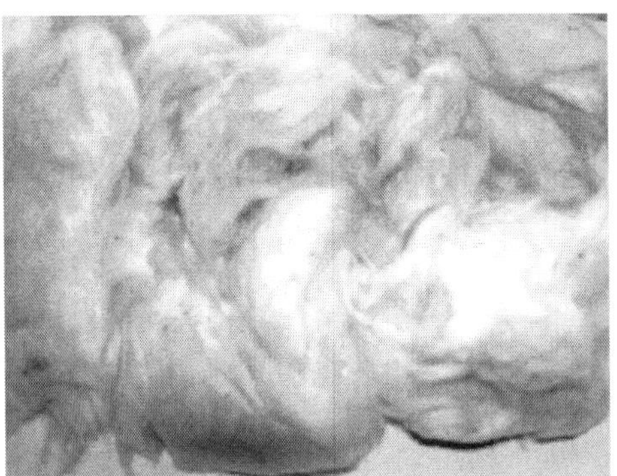

Fig. Kapok

It is used only for filling up mattresses and pillows.

Ramie Fibre

Fig. Ramie Fibre

This is a bast fibre like linen, and is sold as its substitute; It is obtained from a species of nettle plant that grows mainly in Bengal. It also grows in China, Egypt, Java and Japan. The plant grows to a height of 4 to 8 feet. The stalks are cut and water retted, the excessive gum is removed by a chemical. The fibres are combed and straightened.

The long fibres are known as 'line', and the short ones are knownas 'noils'. Linen fibres are spun and woven into a cloth, which is durable and has crispness which makes it roped for attractive table linen. Ramie cloth is known as China or Canton Linen. The noils are used for making canvas *wadding* and cardage.

MATERIALS AND PROCESSING

Quality has two meanings. It is mainly concerned with material and processing that are used in the products. Further it means costly materials or processing or both have been brought into use wish a view to achieve the desired features. High price and quality are generally associated because the particular end-use has stringent needs, which are difficult, and hence expensive, to meet. In view of this relationship, quality in this sense can be referred to as expense quality.

Since the more expensive material and processing tend to give a greater range of properties, this title' has a real meaning. It has, not with standing, certain limitations in that inexpensive materials or processing may produce a product, that is perfectly suited to a special end-use. High or low-expense quality does not imply any judgement on the merits of the product: it relates only to the kind of product. In fact, of course, the objectives is to produce an article that satisfies the demands of the end-use it was designed to serve.

For this cause, it is preferable to, use the term design quality when the merits of a product are judged from the standpoint of end-use suitability. Whether a yarn or fabric has a high end-use suitability relies upon how the design has been translated into reality through production. This introduces the concept of quality measured in the degree of conformity to specification.

Achieving conformance is a production problem and this aspect of quality can usefully be referred to as 'production quality' Circumstances will affect the relative importance of these two aspects. End-use suitability does not merely imply the presence of a particular quality characteristic; rather is it a combination of many separate or related, characteristics or both. In certain cases, once the material has been chosen and the processing sequence decided upon, most of the features are already uniquely determined. In others, the important characteristics may be quite different. In general, however, the characteristics fall into three groups. First, there are those that rely solely on the raw materials and their reactian to acids, heat, etc.

Secondly, there are those qualities that are a reflection of the design and maintenance of the machines utilised and the skill and attension of the operatives attending such machines. For instance, if a roller forming part of the drafting system on a spinning frame is eccentric, the material being processed will contain a periodic wave.

The size and intensity of the wave will rely on the diameter of the roller, the amount of attenuation, and the degree of eccentricity. Finally, there are those that are the combined result of both the material and the processing. For example, in cotton yarn the incidence of neps is mainly dependent on the percentage of immature fibres present.

At the same time, the condition of the carding engine, particularly the state of the clothing and the frequency of stripping and grinding, can also influence the number af neps in the finished yarn. As knowledge of materials and machines becomes greater, the interrelationships divulged increase the importance of this third group.

Meaning

As with quality, there are various opinions as to the meaning of quality control. To some, quality control is only a synonym for testing, whereas others associate it primarily with the application of statistical methods to quality problems. Both views are unnecessarily restrictive.

Broadly, quality control is regarded from two several standpoints; First, it is regarded as the control of all the factors that play a part in determining the quality of the final product. In this sense, it is a basic part of management. Secondly, it is used to describe a series of techniques or procedures that enable management to discharge its quality responsibilities more effectively. To call the latter' statistical quality control' is no solution, since this is concerned only with a few of the many techniques available. Quality control cannot be achieved in a complex textile undertaking without both management and technique. Practically, the deciding factor is that of, management. Effective control depends on managerial policy and action; the techniques, however precise, can serve to improve such action, but they cannot replace it, They spotlight differences between the actual and required standard and, where feasible, measure such differences quantitatively. Control itself depends on the taking of specific action to remedy such differences and this, in turn, involves management.

If the above comments are borne in mind, quality control is defined as being related to all the procedures utilised and action taken to ensure that the quality characteristics of the product conform to the needed standard. This standard may, or may not, be written down, and it may not be especially precise; nor need it cover all the feasibles. The measure of the effectiveness of quality control is then simply the degree to which such conformity is attained.

Standard's Role

Fundamental in quality control is the concept of a standard of quality performance. Here again the word 'standard' is subject to various interpretations. It is important, however, to consider the many meanings in order to provide a basis for a realistic and consistent quality policy. There are at least three-types or aspects of standard that merit consideration. They all play a significant part in establishing a yardstick against which the effectiveness of quality control can be measured.

Standards of Product

The first important standard is the commercial one. This is the quality level desired for the market in which the product is to be sold. In the textile industry the practice is varied. Although cloth constructions are generally laid down, there is considerable variation in the amount of detail on performance characteristics. Specifications of the permissible level of fault incidence likewise vary from the exceptionally detailed to the very vague. For yarnsl it is still common to produce to a specification covering only average count and minimum strengthl. Both these characteristics being judged by a somewhat haphazard series of tests on small samples. It is worth noting, however, that in many sections of the industry the trend is towards more detailed specifications, embodying not only more characteristics but also more systematically defined permissible limits. There are many factors influencing this trend but only a few requirements be mentioned here. First there is the growing importance and influence of integrated manufacturing units.

Controlling as they do part, or all of their processing they admire how various levels of yarn or cloth quality are achieved and how they affect performance at subsequent processing. On those occasions when they supplement their supplies from outside sources, they tend naturally to lay down specifications as similarly detailed as those used within their own units.

Secondly, there is the increasingly important role of the large institutional buyers. Several of these organizations have a direct financial interest in some of the most modern spinning, weaving, and knitting units. Besides, their influence is increased, since they are also the largest customers of some other units.

Laying down detailed quality requirements for the garments they sell, they force the firms at earlier processes to impose similarly stringent demands on their suppliers. Thirdly, one of the factors that has hitherto prevented the preparation of standards on several features is that of inadequate instrumentation. Developments in recent years have enabled more characteristics than previously to be accurately measured and compared with reproducible standards.

Even so, the influence of these forces should not be exaggerated. Many textile firms, in several sections, are still working in a situation where the standards in use are very loosely defined.

Again, where standards exist, they still fluctuate as cyclical trade fluctuations affect the relative bargaining strength of buyers and sellers. This situation has had a fundamental influence on both the level and the control of production quality. Although there are often no clearly laid-down commercial standards, this should not be regarded as a reason for reducing attempts to control quality.

When certain customers are not able to measure quality features in a systematic and precise way, it is incorrect to assume that they do not regard them as important, even though this is not clearly stated. Secondly, several of the quality characteristics are an index not only of the suitability of the product for later use but also of the ease of processing in the producing mill itself. For example, it may be worth while for a spinner to produce a yarn stronger than a particular weaver requires because such a yarn can be processed with fewer breakages at the spinning frame.

With reference to many of the characteristics not closely determined by the raw material used, it is often expensive to process low-quality product. Better product quality is often associated with better and cheaper processing at the producer's mill as well as' at that of the purchaser. Without adequate control procedures and checks, however, it is not feasible for a manufacturer to evaluate the gains made in his own processing. Briefly, a producer is not able to make a consistent decision on what quality to aim for if account is taken only of the demands made by his customers.

Quality Level

Commercial standards relate to the quality level desired for the particular market in which the product is to be sold. More simply, they are those that are referred to by the buyer and seller in relation to their commercial transactions.

Whatever commercial arrangements are made, however, the individual producer should aim at laying down performance standards at each processing stage in as precise and detailed a form as possible. These constitute a second group, which can be called mill quality standards.

For example, to evaluate the setting, scouring, and maintenance procedures at the doorframe, it is essential to measure the continuity of the sliver produced, with regard both to its overall level and to the incidence of periodicities. This could in no way be called a commercial standard, in that no one purchasing the yarn would be directly related to its performance at an early stage of processing.

The objective of mill quality standards is to provide management with a measure of the mill's quality performance. When standards for various quality characteristics are laid down and agreed with the personnel liable for their attainment, management is in a strong position to locate weaknesses in staff, materials, or machines and to take proper action when and where standards are not met. Without objective quality indices, management cannot assess quality performance and its positton is correspondingly weakened. The setting of these standards, which must be realistic, is not easy.

What is realistic in any special situation depends on the inherent variation in the process and on what is possible, bearing in mind the quality of the material or labour employed. To get this information very often requires complex process-capability studies and, as so often, the greatest enemy is wishful thinking.

If the customer is demanding a standard higher than can be met with existing machinery or materials, no useful objective is to be served by arbitrarily tightening mill standards. Commercial and mill standards tend to coincide, in the long run. This is because, in general terms, if the quality is lower that required, the product will not be sold, and if the quality is higher than that desired, it will adversely affect the firm's competitive position because the product will cost more.

In general terms is the key phrase here as the here, because the determination of the standards is taking place in a changing situation, and the actual position sometimes, if ever, reaches equilibrium.

Again, as has been previously stated, it may still pay in certain cases to produce a quality higher than is required if the processing economies that result justify it. The discussion has manufacturing one quality of implicitly assumed that the producer is manufacturing one product. In actual practice, this is rarely so, and often there will be many different quality levels in a mill. It will not always be feasible or desirable to achieve coincident commercial and mill-quality standards, on all of them.

The quality of material and the length and type of processing can be varied, but the mill standards for operative performance and machine condition must be geared to the highest quality being produced. In these respects, long-term quality standards must be the dominant ones, for it is notoriously difficult to restore lowered quality.

Several firms found that they paid dearly for condoning poor operative methods and machine maintenance in the period of lowered commercial standards immediately after the second world war.

Technological and Processing Standards

Eventually, mention should be made of the role of general technological and processing standards. These 'norms' which are Published in different technical journals, as well as in machinery manufacturer specifications, are useful to management despite their invitable generality. They generally relate to what is feasible in the present state of technical knowledge and machine development.

To illustrate, one may quote some of the performance features to be hoped from a modern speed frame employed on the cotton system for coarse-medium counts. The coefficient of variation of weight per unit length within a bobbin should be less than 1 per cent; the Uster regularity percentage should not cross 3.5 per cent and there should be no periodicities; it should be possible to change the staple length processed by 3/16 in.

Without changing settings, with a maximum deterioration of 0.5 per cent. This information is useful in helping management to evaluate the feasible gain to be made in quality in relation to the capital expenditure necessary to achieve it. Besides, it gives an indication of the hoped future level of competition in quality.

Unfortunately however, it does not provide much assistance to a Particular firm in fixing its mill-quality standards. Most firms are more interested in finding out how they compare. With other firms utilising similar processing and machines of comparable age, rather than with some theoretical ideal.

Such information is rarely available, and there is a tendency for managements to attribute all the quality differences to better machinery or various processing. Actual inter-firm comparison are so infrequent, and information is so scanty, that it is difficult to know if mill-quality standards are realistic. There is undoubtedly a great deal of confusion over standards in many textile. firms.

This generally results in an unwillingness or inability to lay down what is to be achieved at the many stages of manufacture. It is important that the scope and aim in any particular case should be laid down and clearly understood by all concerned. Only then is it feasible to devise a quality policy to meet these aims and, by measuring the results of such a policy, to evaluate its effectiveness.

There are many difficulties in drawing up the many standards that form an inevitable part of a quality policy. One approach, which the author has found to be of value in widely differing circumstances, is that of 'quality audit'. It is not possible here to give a detailed description, and a brief outline only is included below. A quality audit can be regarded as a logical application of the scientific method to the problem, of qua!ity improvement.

In the beginning, information is collected on all those factors that affect the final quality of the product. The central feature of this initial stage is the series of observations carried out by a team drawn from the mill staff. These obseryations give details of the processing conditions on the various machines and the methods employed by the operatives attending those machines. Concurrently, inspections are carried out to assess the mechanical state of the machines in question and to measure sure the environmental conditions in which processing is being undertaken.

To complete the picture, the material itself is tested at various stages of processing. This provides more information on the effects of these various factors on the quality of the final product. These observations, inspections, and tests are planned in such a way that it is feasible to interpret and analyse the, interrelationships that they reveal by the use of nothing more than the simplest statistical methods. This analysis forms the second stage of the audit, the object being to trace back. important quality differences to the relevant technical and operating conditions.

From the results of the analysis a report is drawn up, this providing a complete account of the quality position of the mill. While the various kinds of remedial action are being undertaken, routine' control methods are planned and installed to make sure that the improvements secured are rnaintained. There are four main features of the approach that merit emphasis.

First, the audit deals with all those factors that affect the final quality of the product and is not confined to the normal laboratory tests, which give only part of the picture. The common difficulty in doing this is that there are so many factors that the data become unmanageable. The problem is alleviated in this method by building the data around investigation of one quality indicator. For spinning, doubling, and weaving, the number of 'end-down' is used.

It is naturall much easier to study interrelationships betweeen different factors if some sort of framework is available. The position is also eased by the use of very simple-methods of quality assessment, live the use of the +; ±; - scale to evaluate certain characteristics. Obviously, other and more precise methods are employed, but generally the simple methods are sufficient to reveal processing or technical weaknesses.

The problem is the general one of trying to comply with the aims, which are sometimes mutually inconsistent, of keeping the data down to manageable proportions, getting them to be sufficiently accurate to be significant, and doing the job properly quickly with the kind of observer available. Since data are collected on all the factors affecting quality, the ensuing report thus provides a complete account of the quality position at one point in time. The interrelationship and interdependence of the various factors are clearly revealed, and such inconsistencies as may be present are highlighted.

Managements are therefore able to see the entire picture and can then use this to establish a quality policy. Secondly, the audit brings together the different groups of people who have. special responsibility for quality control. All learn to understand the contribution made by each other. In particular, the co-operation between laboratory and mill staff encouraged by the audit is likeiy to be continued afterwards. Many of the data-collection methods used can be adapted by the quality-control staff to help the supervisor to keep a check on the quality aspects of machine and operative performance.

The third feature is that of carrying out the whole survey with the mill's own staff. This helps both in the gathering of data and in the acceptance of results. Since all levels of mill staff have played some part in the final report, there is much greater likelihood that action to remedy particular faults win be taken. Inevitably, the approach or method is aimed at helping the mill to help itself. Finally, the mill is being compared with itself. Although technological standards are used for comparison with regard to such characteristics as count, irregularity, etc., on many aspects none are available. In any case, to use such standards would be very unrealistic, especially when an older mill is being considered. Mill standards, however, are always computed.

For example, if one group of frames is significantly worse than the rest in some respect, then attention can be directed to those frames, even though there are no 'technological standards' available for frames of that age and condition. Even if the long-term solution lies in re-equipment, some standards are essential to ensure that an the frames are brought to a common level, however low this might be.

Control's Framework

In a publication as general as this, the intention is not to describe particular control procedures but rather to provide some indication of the anatomy of control. Furthermore, although the many factors are interrelated, for ease of presentation they have been considred under three headings: Materials Machines, and Operatives.

Materials

In the beginning, it is essential to check whether the raw material, yarn, or cloth purchased conforms to the specification laid down. The significance of this aspect of control varies markedly from one section of the industry to another. For example, in cotton spinning it is feasible and desirable' to make on the raw cotton a range of tests covering such features as staple length, maturity, percentage short fibre, fibre strength and fineness, and trash content. The whole of the subsequent processing will be affected by the results of such tests. On the other hand, the weaver is often in the position of being unable to carry out more than the most rudimentary check on the standard of the incoming yarn, especially where beams are concerned.

Unless the small sample reveals a major fault, the usual objective is not so much to make between good or bad batches put, over a period of time, to select good suppliers. Even apart from the sample-size problem, there is the difficulty of the characteristics to be tested.

This is particularly so in weaving where the factors that determine the performance of a certain set in the loom are just as likely to be found in the way it was prepared as in the inherent qualities of the yarn. After the initial raw-material testing, the material may be tested after or during subsequent processing. These tests are to give both a check on the running of the process and a basis for action where results differ from the standard. These checks are an important part of control in such process industries as spinning, rather than in assembly ones such as weaving or knitting. For example, the regularity of a drawframe sliver may be tested on a daily basis. The over-all level provides an indication of the processing up to and including the drawframe process.

The existence of certain periodicities of varying wave lengths also provides an indication of the areas where action needs to be taken. In dealing with the final stage of material testing, *i.e.*, the outgoing finished product, the aims of the procedures employed are three-fold. First they serve to give a check on the characteristics of the product to see whether they similar to customer standards. The check may result in the despatch of the cloth or yarn, etc., as satisfactory or it may lead to its being withheld.

Between these two extremes, there are an infinite number of possible courses of action, which may comprise price allowances. Where the product is supplied to customers with varying acceptance standards, the final testing may be used to provide a basis for allotting the product to the different customers.

Secondly, the degree of conformance of the final product to the standard laid down may be regarded as an indication of the effectiveness of quality control in the mill; for instance, with such features as the variation in count, which reflects the eventual effectiveness of a series of controls at various stages of processing. Thirdly, the testing of the product can be regarded as a check on the final process.

The regularity of yarn, for instance, taken in conjunction with that of the roving fed to the ringframe, connotes not only an important feature of the final product but also the degree of control in the ringroom.

Controlling Procedures

Theoretically, were it feasible to setup clearly the exact causes of variations from standard by examining the product, there would be no need to consider control procedures other than those related to material and product testing. Practically, however, this is not feasible, and, even if it were, it does not necessarily follow that this would be the best thing to do. This is because the type of control procedure affects the likelihood that action will be taken on the results. For example, if yarn is examined and periodicities are revealed, it is feasible to conclude that an eccentric roller is causing the fault.

At the same time, there may be a multitude of causes that have contributed, and the most direct way is to examine the rollers with an eccentricity gauge. This has the added benefit that it can be performed by the overlooker himself. Regarding machines, it can be argued that it is first essential to check whether the postulated quality characteristics are consistent with the capabilities of the machine.

This corresponds to the problem of selecting materials that can be expected to give certain characteristics in the final product. For example, just as it is impossible to spin yarns of a certain count and strength from some cottons, it is equally impossible to produce yarn of a certain level of regularity from machines of certain age and design.

There is also the question of whether it is feasible to produce a some quality by processing the material through a limited sequence of machines. These last two factors are interrelated with the process capabilities of the individual machines making up that sequence.

The action associated with these checks and process-capability studies will depend on both the results and the time factor. Small deviations from standard may lead to adjustment in the succeeding process or to changing the feeding procedures at the machine tested.

Machine checks initially comprise ensuring that the machine has been correctly set up, *i.e.*, that the speeds and settings are consistent with obtaining the desired characteristics in the product. It is also essential to check periodically that these speeds and settings remain as laid down. Checks should be made to ensure that the condition of the machines is being maintained at a level consistent with quality needs. This provides a good example of how various levels of management are affected by the results of a particular control procedure.

It may divulge the need for relatively minor action, e. g., the fitting of new traveller cleaners in ring-spinning frames. The decision here can be made at departmental level. It may reveal the necessity for changes in the stipulated maintenance programme, *e.g.*., increasing scouring frequencies. The proper action here involves. the mill manager in that the effects of such a change will affect the balance of production in all departments.

Quality Control Operatives

Regarding operatives, there are really three chief areas of quality-control checks or procedures. Each must be covered to make sure that quality reaches, and is maintained at, the required level. First, it is necessary to establish whether the operatives possess the level of ability needed.

Associated closely with this aspect of control, which is more accurately part of selection, is the question of training the operatives once they have been opted, It is very easy to do this with new operatives if systematic training procedures are adopted, but much more difficult with experienced Operatives whose bad quality habits are deeply rooted.

Quality practices are by no means standardized, and it follows that each operative should be clearly informed of what he or she is supposed to do in very considerable detail. This can be discussed by the possible ways of carrying out the apparently straightforward task of creeling in a can at the drawframe. There are two normal ways of doing this:

- Break the sliver and piece it to the beginning of a new can, a remnant being left in the bottom of the old one.
- Piece start or the full can to the end of the preceding can; this comprises removing the remainder of the sliver from the old can while the frame is running, turning it over, and piecing the end to the beginning of the full can; and

The piece that is created as an out come of the adoption of the second method can, in turn, be disposed of in one of several ways:

- It can be placed in a special 'piece can'; or
- It can be pieced to the top of the full can later;
- The can comprising it can be transferred to the front of the preceding head and used instead of an empty can at doff.
- It can be discarded as, vaste;

The quality of the product is explicitly affected by the creation and use of these pieces. If the first or third means of disposal is employed, two further piecings are necessary to put the piece back into the system.

The 'piece-can' method has an additional and obvious detrimental effect on quality. The last method needs only one join to put a piece back into the system, but, if the practice is continued, the piece,will remain in the can for a long period of time. This particular aspect has been given in some detail to illustrate the fact that there are often several ways of performing a relatively straight forward task in an individual situation.

To make sure that the correct one is used involves not only a detailed breakdown of the task but clear instructions on each part of it, followed by regular checks to ensure that the stipulated procedure is adhered to. The second area requiring attention is the work allocation. Does it allow the achievement of the quality standards? This must be checked, for it cannot be assumed to be correct.

In this regard, the role of techniques such as work study is important. For instance, it is just as significant to calculate the amount of cleaning necessary as it is to discover the amount of time it takes to clean. Too often cleaning frequencies bear no relation to the standard of cleanliness of machines required for a given level of product cleanliness. The work-study officer is not liable for the decision on such frequencies: he merely takes them into account in calculating the over-all work-load.

The precision in timing a special operation is likely, however, to give an aura of objectivity, which conceals the fact that decisions on the nature and extent of cleaning are frequently made quite unsystematically. If one considers work organization in the broader aspect, there seems little doubt that some systems of work allocation do adversely affect the quality of the product. For instance, it is customary in ring spinning for the operative to be for patrolling the spinning frames both to repair broken ends and to replenish the empty supply packages.

Some work-loading systems have aimed at separating these two jobs. Theoretically, there should be no overlap between the duties allocated to each group. In practice it is by no means so simple. The division of responsibilities tends to have an adverse effect on quality, with respect both to the individual operative, who finds other people are playing a part in affecting the processing conditions, and to the supervisor, who finds it difficult to allocate responsibility. This is not necessarily to imply that these disadvantages may not be more than offset by increased machine efficiency, etc.

In general terms then, it is wrong to consider standards of operative performance without regard to the particular work organization in which they are expected to operate. This should be interpreted in the widest sense and should not be confined to the division of duties.

8

Optical Fibre

An optical fibre (or optical fibre) is a flexible, transparent fibre made of extruded glass (silica) or plastic, slightly thicker than a human hair. It can function as a waveguide, or "light pipe", to transmit light between the two ends of the fibre. The field of applied science and engineering concerned with the design and application of optical fibers is known as fibre optics.

Optical fibers are widely used in fibre-optic communications, where they permit transmission over longer distances and at higher bandwidths (data rates) than wire cables. Fibers are used instead of metal wires because signals travel along them with less loss and are also immune to electromagnetic interference. Fibers are also used for illumination, and are wrapped in bundles so that they may be used to carry images, thus allowing viewing in confined spaces. Specially designed fibers are used for a variety of other applications, including sensors and fibre lasers. Optical fibers typically include a transparent core surrounded by a transparent cladding material with a lower index of refraction. Light is kept in the core by total internal reflection. This causes the fibre to act as a waveguide. Fibers that support many propagation paths or transverse modes are called multi-mode fibers (MMF), while those that only support a single mode are called single-mode fibers (SMF). Multi-mode fibers generally have a wider core diameter, and are used for short-distance communication links and for applications where high power must be transmitted. Single-mode fibers are used for most communication links longer than 1,000 meters (3,300 ft).

Joining lengths of optical fibre is more complex than joining electrical wire or cable. The ends of the fibers must be carefully cleaved, and then carefully spliced together with the cores perfectly aligned. A mechanical splice holds the ends of the fibers together mechanically, while fusion splicing uses heat to fuse the ends of the fibers together. Special optical fibre connectors for temporary or semi-permanent connections are also available.

HISTORY

Guiding of light by refraction, the principle that makes fibre optics possible, was first demonstrated by Daniel Colladon and Jacques Babinet in Paris in the early 1840s. John Tyndall included a demonstration of it in his public lectures in London, 12 years later. Tyndall also wrote about the property of total internal reflection in an introductory book about the nature of light in 1870:

- When the light passes from air into water, the refracted ray is bent *towards* the perpendicular... When the ray passes from water to air it is bent *from* the perpendicular... If the angle which the ray in water encloses with the perpendicular to the surface be greater than 48 degrees, the ray will not quit the water at all: it will be *totally reflected* at the surface.... The angle which marks the limit where total reflection begins is called the limiting angle of the medium. For water this angle is 48°27', for flint glass it is 38°41', while for diamond it is 23°42'.

Unpigmented human hairs have also been shown to act as an optical fibre.

Practical applications, such as close internal illumination during dentistry, appeared early in the twentieth century. Image transmission through tubes was demonstrated independently by the radio experimenter Clarence Hansell and the television pioneer John Logie Baird in the 1920s. The principle was first used for internal medical examinations by Heinrich Lamm in the following decade. Modern optical fibers, where the glass fibre is coated with a transparent cladding to offer a more suitable refractive index, appeared later in the decade. Development then focused on fibre bundles for image transmission. Harold Hopkins and Narinder Singh Kapany at Imperial College in London achieved low-loss light transmission through a 75 cm long bundle which combined several thousand fibers.

Their article titled “A flexible fibrescope, using static scanning” was published in the journal *Nature* in 1954. The first fibre optic semi-flexible gastroscope was patented by Basil Hirschowitz, C. Wilbur Peters, and Lawrence E. Curtiss, researchers at the University of Michigan, in 1956. In the process of developing the gastroscope, Curtiss produced the first glass-clad fibers; previous optical fibers had relied on air or impractical oils and waxes as the low-index cladding material.

A variety of other image transmission applications soon followed.

In 1880 Alexander Graham Bell and Sumner Tainter invented the Photophone at the Volta Laboratory in Washington, D.C., to transmit voice signals over an optical beam. It was an advanced form of telecommunications, but subject to atmospheric interferences and impractical until the secure transport of light that would be offered by fibre-optical systems. In the late 19th and early 20th centuries, light was guided through bent glass rods to illuminate body cavities. Jun-ichi Nishizawa, a Japanese scientist at Tohoku University, also proposed the use of optical fibers for communications in 1963, as stated in his book published in 2004 in India. Nishizawa invented other technologies that contributed to the development of optical fibre communications, such as the graded-index optical fibre as a channel for transmitting light from semiconductor lasers. The first working fibre-optical data transmission system was demonstrated by German physicist Manfred Börner at Telefunken Research Labs in Ulm in 1965, which was followed by the first patent application for this technology in 1966. Charles K. Kao and George A. Hockham of the British company Standard Telephones and Cables (STC) were the first to promote the idea that the attenuation in optical fibers could be reduced below 20 decibels per kilometer (dB/km), making fibers a practical communication medium. They proposed that the attenuation in fibers available at the time was caused by impurities that could be removed, rather than by fundamental physical effects such as scattering. They correctly and systematically theorized the light-loss properties for optical fibre, and pointed out the right material to use for such fibers — silica glass with high purity. This discovery earned Kao the Nobel Prize in Physics in 2009.

NASA used fibre optics in the television cameras that were sent to the moon. At the time, the use in the cameras was classified *confidential*, and only those with sufficient security clearance or those accompanied by someone with the right security clearance were permitted to handle the cameras.

The crucial attenuation limit of 20 dB/km was first achieved in 1970, by researchers Robert D. Maurer, Donald Keck, Peter C. Schultz, and Frank Zimar working for American glass maker Corning Glass Works, now Corning Incorporated. They demonstrated a fibre with 17 dB/km attenuation by doping silica glass with titanium. A few years later they produced a fibre with only 4 dB/km attenuation using germanium dioxide as the core dopant. Such low attenuation ushered in the era of optical fibre telecommunication. In 1981, General Electric produced fused quartz ingots that could be drawn into strands 25 miles (40 km) long.

Attenuation in modern optical cables is far less than in electrical copper cables, leading to long-haul fibre connections with repeater distances of 70–150 kilometers (43–93 mi). The erbium-doped fibre amplifier, which reduced the cost of long-distance fibre systems by reducing or eliminating optical-electrical-optical repeaters, was co-developed by teams led by David N. Payne of the University of Southampton and Emmanuel Desurvire at Bell Labs in 1986. Robust modern optical fibre uses glass for both core and sheath, and is therefore less prone to aging. It was invented by Gerhard Bernsee of Schott Glass in Germany in 1973.

The emerging field of photonic crystals led to the development in 1991 of photonic-crystal fibre, which guides light by diffraction from a periodic structure, rather than by total internal reflection. The first photonic crystal fibers became commercially available in 2000. Photonic crystal fibers can carry higher power than conventional fibers and their wavelength-dependent properties can be manipulated to improve performance.

USES

FIBRE-OPTIC COMMUNICATION

Fibre-optic communication is a method of transmitting information from one place to another by sending pulses of light through an optical fibre. The light forms an electromagnetic carrier wave that is modulated to carry information. First developed in the 1970s, fibre-optic communication systems have revolutionized the telecommunications industry and have played a major role in the advent of the Information Age. Because of its advantages over electrical transmission, optical fibers have largely replaced copper wire communications in core networks in the developed world. Optical fibre is used by many telecommunications companies to transmit telephone signals, Internet communication, and cable television signals. Researchers at Bell Labs have reached internet speeds of over 100 petabits per second using fibre-optic communication.

The process of communicating using fibre-optics involves the following basic steps: Creating the optical signal involving the use of a transmitter, relaying the signal along the fibre, ensuring that the signal does not become too distorted or weak, receiving the optical signal, and converting it into an electrical signal.

Applications

Optical fibre is used by many telecommunications companies to transmit telephone signals, Internet communication, and cable television signals. Due to much lower attenuation and interference, optical fibre has large advantages over existing copper wire in long-distance and high-demand applications. However, infrastructure development within cities was relatively difficult and time-consuming, and fibre-optic systems were complex and expensive to install and operate.

Due to these difficulties, fibre-optic communication systems have primarily been installed in long-distance applications, where they can be used to their full transmission capacity, offsetting the increased cost. Since 2000, the prices for fibre-optic communications have dropped considerably. The price for rolling out fibre to the home has currently become more cost-effective than that of rolling out a copper based network. Prices have dropped to $850 per subscriber in the US and lower in countries like The Netherlands, where digging costs are low and housing density is high.

Since 1990, when optical-amplification systems became commercially available, the telecommunications industry has laid a vast network of intercity and transoceanic fibre communication lines. By 2002, an intercontinental network of 250,000 km of submarine communications cable with a capacity of 2.56 Tb/s was completed, and although specific network capacities are privileged information, telecommunications investment reports indicate that network capacity has increased dramatically since 2004.

History

In 1880 Alexander Graham Bell and his assistant Charles Sumner Tainter created a very early precursor to fibre-optic communications, the Photophone, at Bell's newly established Volta Laboratory in Washington, D.C. Bell considered it his most important invention. The device allowed for the transmission of sound on a beam of light. On June 3, 1880, Bell conducted the world's first wireless telephone transmission between two buildings, some 213 meters apart. Due to its use of an atmospheric transmission medium, the Photophone would not prove practical until advances in laser and optical fibre technologies permitted the secure transport of light. The Photophone's first practical use came in military communication systems many decades later.

In 1966 Charles K. Kao and George Hockham proposed optical fibers at STC Laboratories (STL) at Harlow, England, when they showed that the losses of 1000 dB/km in existing glass (compared to 5-10 dB/km in coaxial cable) was due to contaminants, which could potentially be removed.

Optical fibre was successfully developed in 1970 by Corning Glass Works, with attenuation low enough for communication purposes (about 20dB/km), and at the same time GaAs semiconductor lasers were developed that were compact and therefore suitable for transmitting light through fibre optic cables for long distances.

After a period of research starting from 1975, the first commercial fibre-optic communications system was developed, which operated at a wavelength around 0.8 μm and used GaAs semiconductor lasers. This first-generation system operated at a bit rate of 45 Mbps with repeater spacing of up to 10 km. Soon on 22 April 1977, General Telephone and Electronics sent the first live telephone traffic through fibre optics at a 6 Mbit/s throughput in Long Beach, California.

The first wide area network fibre optic cable system in the world seems to have been installed by Rediffusion in Hastings, East Sussex, UK in 1978. The cables were placed in ducting throughout the town, and had over 1000 subscribers. They were used at that time for the transmission of television channels,not available because of local reception problems. The system is still in place, but disused.

The second generation of fibre-optic communication was developed for commercial use in the early 1980s, operated at 1.3 μm, and used InGaAsP semiconductor lasers. These early systems were initially limited by multi mode fibre dispersion, and in 1981 the single-mode fibre was revealed to greatly improve system performance, however practical connectors capable of working with single mode fibre proved difficult to develop. By 1987, these systems were operating at bit rates of up to 1.7 Gb/s with repeater spacing up to 50 km.

The first transatlantic telephone cable to use optical fibre was TAT-8, based on Desurvire optimized laser amplification technology. It went into operation in 1988.

Third-generation fibre-optic systems operated at 1.55 μm and had losses of about 0.2 dB/km. This development was spurred by the discovery of Indium gallium arsenide and the development of the Indium Gallium Arsenide photodiode by Pearsall. Engineers overcame earlier difficulties with pulse-spreading at that wavelength using conventional InGaAsP semiconductor lasers. Scientists overcame this difficulty by using dispersion-shifted fibers designed to have minimal dispersion at 1.55 μm or by limiting the laser spectrum to a single longitudinal mode. These developments eventually allowed third-generation systems to operate commercially at 2.5 Gbit/s with repeater spacing in excess of 100 km.

The fourth generation of fibre-optic communication systems used optical amplification to reduce the need for repeaters and wavelength-division multiplexing to increase data capacity. These two improvements caused a revolution that resulted in the doubling of system capacity every 6 months starting in 1992 until a bit rate of 10 Tb/s was reached by 2001. In 2006 a bit-rate of 14 Tbit/s was reached over a single 160 km line using optical amplifiers.

The focus of development for the fifth generation of fibre-optic communications is on extending the wavelength range over which a WDM system can operate. The conventional wavelength window, known as the C band, covers the wavelength range 1.53-1.57 μm, and *dry fibre* has a low-loss window promising an extension of that range to 1.30-1.65 μm. Other developments include the concept of "optical solitons, " pulses that preserve their shape by counteracting the effects of dispersion with the nonlinear effects of the fibre by using pulses of a specific shape.

In the late 1990s through 2000, industry promoters, and research companies such as KMI, and RHK predicted massive increases in demand for communications bandwidth due to increased use of the Internet, and commercialization of various bandwidth-intensive consumer services, such as video on demand. Internet protocol data traffic was increasing exponentially, at a faster rate than integrated circuit complexity had increased under Moore's Law. From the bust of the dot-com bubble through 2006, however, the main trend in the industry has been consolidation of firms and offshoring of manufacturing to reduce costs. Companies such as Verizon and AT and T have taken advantage of fibre-optic communications to deliver a variety of high-throughput data and broadband services to consumers' homes.

Technology

Modern fibre-optic communication systems generally include an optical transmitter to convert an electrical signal into an optical signal to send into the optical fibre, a cable containing bundles of multiple optical fibers that is routed through underground conduits and buildings, multiple kinds of amplifiers, and an optical receiver to recover the signal as an electrical signal. The information transmitted is typically digital information generated by computers, telephone systems, and cable television companies.

Transmitters

The most commonly used optical transmitters are semiconductor devices such as light-emitting diodes (LEDs) and laser diodes. The difference between LEDs and laser diodes is that LEDs produce incoherent light, while laser diodes produce coherent light. For use in optical communications, semiconductor optical transmitters must be designed to be compact, efficient, and reliable, while operating in an optimal wavelength range, and directly modulated at high frequencies.

In its simplest form, a LED is a forward-biased p-n junction, emitting light through spontaneous emission, a phenomenon referred to as electroluminescence. The emitted light is incoherent with a relatively wide spectral width of 30-60 nm. LED light transmission is also inefficient, with only about 1 per cent of input power, or about 100 microwatts, eventually converted into launched power which has been coupled into the optical fibre. However, due to their relatively simple design, LEDs are very useful for low-cost applications.

Communications LEDs are most commonly made from Indium gallium arsenide phosphide (InGaAsP) or gallium arsenide (GaAs). Because InGaAsP LEDs operate at a longer wavelength than GaAs LEDs (1.3 micrometers vs. 0.81-0.87 micrometers), their output spectrum, while equivalent in energy is wider in wavelength terms by a factor of about 1.7. The large spectrum width of LEDs is subject to higher fibre dispersion, considerably limiting their bit rate-distance product (a common measure of usefulness). LEDs are suitable primarily for local-area-network applications with bit rates of 10-100 Mbit/s and transmission distances of a few kilometers. LEDs have also been developed that use several quantum wells to emit light at different wavelengths over a broad spectrum, and are currently in use for local-area WDM (Wavelength-Division Multiplexing) networks.

Today, LEDs have been largely superseded by VCSEL (Vertical Cavity Surface Emitting Laser) devices, which offer improved speed, power and spectral properties, at a similar cost. Common VCSEL devices couple well to multi mode fibre.

A semiconductor laser emits light through stimulated emission rather than spontaneous emission, which results in high output power (~100 mW) as well as other benefits related to the nature of coherent light. The output of a laser is relatively directional, allowing high coupling efficiency (~50 per cent) into single-mode fibre. The narrow spectral width also allows for high bit rates since it reduces the effect of chromatic dispersion. Furthermore, semiconductor lasers can be modulated directly at high frequencies because of short recombination time.

Commonly used classes of semiconductor laser transmitters used in fibre optics include VCSEL (Vertical-Cavity Surface-Emitting Laser), Fabry–Pérot and DFB (Distributed Feed Back).

Laser diodes are often directly modulated, that is the light output is controlled by a current applied directly to the device. For very high data rates or very long distance *links*, a laser source may be operated continuous wave, and the light modulated by an external device such as an electro-absorption modulator or Mach–Zehnder interferometer. External modulation increases the achievable link distance by eliminating laser chirp, which broadens the linewidth of directly modulated lasers, increasing the chromatic dispersion in the fibre.

A transceiver is a device combining a transmitter and a receiver in a single housing.

Receivers

The main component of an optical receiver is a photodetector, which converts light into electricity using the photoelectric effect. The primary photodetectors for telecommunications are made from Indium gallium arsenide The photodetector is typically a semiconductor-based photodiode. Several types of photodiodes include p-n photodiodes, p-i-n photodiodes, and avalanche photodiodes. Metal-semiconductor-metal (MSM) photodetectors are also used due to their suitability for circuit integration in regenerators and wavelength-division multiplexers.

Optical-electrical converters are typically coupled with a transimpedance amplifier and a limiting amplifier to produce a digital signal in the electrical domain from the incoming optical signal, which may be attenuated and distorted while passing through the channel. Further signal processing such as clock recovery from data (CDR) performed by a phase-locked loop may also be applied before the data is passed on.

Fibre cable types

An optical fibre cable consists of a core, cladding, and a buffer (a protective outer coating), in which the cladding guides the light along the core by using the method of total internal reflection. The core and the cladding (which has a lower-refractive-index) are usually made of high-quality silica glass, although they can both be made of plastic as well. Connecting two optical fibers is done by fusion splicing or mechanical splicing and requires special skills and interconnection technology due to the microscopic precision required to align the fibre cores.

Two main types of optical fibre used in optic communications include multi-mode optical fibers and single-mode optical fibers. A multi-mode optical fibre has a larger core (e" 50 micrometers), allowing less precise, cheaper transmitters and receivers to connect to it as well as cheaper connectors. However, a multi-mode fibre introduces multimode distortion, which often limits the bandwidth and length of the link. Furthermore, because of its higher dopant content, multi-mode fibers are usually expensive and exhibit higher attenuation. The core of a single-mode fibre is smaller (<10 micrometers) and requires more expensive components and interconnection methods, but allows much longer, higher-performance links.

In order to package fibre into a commercially viable product, it typically is protectively coated by using ultraviolet (UV), light-cured acrylate polymers, then terminated with optical fibre connectors, and finally assembled into a cable. After that, it can be laid in the ground and then run through the walls of a building and deployed aerially in a manner similar to copper cables. These fibers require less maintenance than common twisted pair wires, once they are deployed.

Specialized cables are used for long distance subsea data transmission, *e.g.*. transatlantic communications cable. New (2011–2013) cables operated by commercial enterprises (Emerald Atlantis, Hibernia Atlantic) typically have four strands of fibre and cross the Atlantic (NYC-London) in 60-70ms.

Another common practice is to bundle many fibre optic strands within long-distance power transmission cable. This exploits power transmission rights of way effectively, ensures a power company can own and control the fibre required to monitor its own devices and lines, is effectively immune to tampering, and simplifies the deployment of smart grid technology.

Amplifier

The transmission distance of a fibre-optic communication system has traditionally been limited by fibre attenuation and by fibre distortion. By using opto-electronic repeaters, these problems have been eliminated. These repeaters convert the signal into an electrical signal, and then use a transmitter to send the signal again at a higher intensity than was received, thus counteracting the loss incurred in the previous segment. Because of the high complexity with modern wavelength-division multiplexed signals (including the fact that they had to be installed about once every 20 km), the cost of these repeaters is very high.

An alternative approach is to use an optical amplifier, which amplifies the optical signal directly without having to convert the signal into the electrical domain. It is made by doping a length of fibre with the rare-earth mineral erbium, and *pumping* it with light from a laser with a shorter wavelength than the communications signal (typically 980 nm). Amplifiers have largely replaced repeaters in new installations.

Wavelength-division multiplexing

Wavelength-division multiplexing (WDM) is the practice of multiplying the available capacity of optical fibers through use of parallel channels, each channel on a dedicated wavelength of light. This requires a wavelength division multiplexer in the transmitting equipment and a demultiplexer (essentially a spectrometer) in the receiving equipment. Arrayed waveguide gratings are commonly used for multiplexing and demultiplexing in WDM. Using WDM technology now commercially available, the bandwidth of a fibre can be divided into as many as 160 channels to support a combined bit rate in the range of 1.6 Tbit/s.

Parameters

Bandwidth–distance product

Because the effect of dispersion increases with the length of the fibre, a fibre transmission system is often characterized by its *bandwidth–distance product*, usually expressed in units of MHz·km. This value is a product of bandwidth and distance because there is a trade off between the bandwidth of the signal and the distance it can be carried. For example, a common multi-mode fibre with bandwidth–distance product of 500 MHz·km could carry a 500 MHz signal for 1 km or a 1000 MHz signal for 0.5 km. Engineers are always looking at current limitations in order to improve fibre-optic communication, and several of these restrictions are currently being researched.

Record speeds

Each fibre can carry many independent channels, each using a different wavelength of light (wavelength-division multiplexing). The net data rate (data rate without overhead bytes) per fibre is the per-channel data rate reduced by the FEC overhead, multiplied by the number of channels (usually up to eighty in commercial dense WDM systems as of 2008).

Year	Organization	Effective speed	WDM channels	Per channel speed	Distance
2009	Alcatel-Lucent	15 Tbit/s	155	100 Gbit/s	90 km
2010	NTT	69.1 Tbit/s	432	171 Gbit/s	240 km
2011	KIT	26 Tbit/s	1	26 Tbit/s	50 km
2011	NEC	101 Tbit/s	370	273 Gbit/s	165 km
2012	NEC, Corning	1.05 Petabit/s	12 core fiber		52.4 km

While the physical limitations of electrical cable prevent speeds in excess of 10 Gigabits per second, the physical limitations of fibre optics have not yet been reached. In 2013, *New Scientist* reported that a team at the University of Southampton had achieved a throughput of 73.7 Tbit per second, with the signal traveling at 99.7 per cent the speed of light through a hollow-core photonic crystal fibre.

Dispersion

For modern glass optical fibre, the maximum transmission distance is limited not by direct material absorption but by several types of dispersion, or spreading of optical pulses as they travel along the fibre. Dispersion in optical fibers is caused by a variety of factors. Intermodal dispersion, caused by the different axial speeds of different transverse modes, limits the performance of multi-mode fibre. Because single-mode fibre supports only one transverse mode, intermodal dispersion is eliminated.

In single-mode fibre performance is primarily limited by chromatic dispersion (also called group velocity dispersion), which occurs because the index of the glass varies slightly depending on the wavelength of the light, and light from real optical transmitters necessarily has nonzero spectral width (due to modulation). Polarization mode dispersion, another source of limitation, occurs because although the single-mode fibre can sustain only one transverse mode, it can carry this mode with two different polarizations, and slight imperfections or distortions in a fibre can alter the propagation velocities for the two polarizations. This phenomenon is called fibre birefringence and can be counteracted by polarization-maintaining optical fibre. Dispersion limits the bandwidth of the fibre because the spreading optical pulse limits the rate that pulses can follow one another on the fibre and still be distinguishable at the receiver.

Some dispersion, notably chromatic dispersion, can be removed by a 'dispersion compensator'. This works by using a specially prepared length of fibre that has the opposite dispersion to that induced by the transmission fibre, and this sharpens the pulse so that it can be correctly decoded by the electronics.

Attenuation

Fibre attenuation, which necessitates the use of amplification systems, is caused by a combination of material absorption, Rayleigh scattering, Mie scattering, and connection losses. Although material absorption for pure silica is only around 0.03 dB/km (modern fibre has attenuation around 0.3 dB/km), impurities in the original optical fibers caused attenuation of about 1000 dB/km. Other forms of attenuation are caused by physical stresses to the fibre, microscopic fluctuations in density, and imperfect splicing techniques.

Transmission windows

Each effect that contributes to attenuation and dispersion depends on the optical wavelength. The wavelength bands (or windows) that exist where these effects are weakest are the most favorable for transmission. These windows have been standardized, and the currently defined bands are the following:

Band	Description	Wavelength Range
O band	original	1260 to 1360 nm
E band	extended	1360 to 1460 nm
S band	short wavelengths	1460 to 1530 nm
C band	conventional ("erbium window")	1530 to 1565 nm
L band	long wavelengths	1565 to 1625 nm
U band	ultralong wavelengths	1625 to 1675 nm

Note that this table shows that current technology has managed to bridge the second and third windows that were originally disjoint.

Historically, there was a window used below the O band, called the first window, at 800-900 nm; however, losses are high in this region so this window is used primarily for short-distance communications. The current lower windows (O and E) around 1300 nm have much lower losses. This region has zero dispersion. The middle windows (S and C) around 1500 nm are the most widely used. This region has the lowest attenuation losses and achieves the longest range. It does have some dispersion, so dispersion compensator devices are used to remove this.

Regeneration

When a communications link must span a larger distance than existing fibre-optic technology is capable of, the signal must be *regenerated* at intermediate points in the link by optical communications repeaters. Repeaters add substantial cost to a communication system, and so system designers attempt to minimize their use.

Recent advances in fibre and optical communications technology have reduced signal degradation so far that *regeneration* of the optical signal is only needed over distances of hundreds of kilometers. This has greatly reduced the cost of optical networking, particularly over undersea spans where the cost and reliability of repeaters is one of the key factors determining the performance of the whole cable system. The main advances contributing to these performance improvements are dispersion management, which seeks to balance the effects of dispersion against non-linearity; and solitons, which use nonlinear effects in the fibre to enable dispersion-free propagation over long distances.

Last mile

Although fibre-optic systems excel in high-bandwidth applications, optical fibre has been slow to achieve its goal of fibre to the premises or to solve the last mile problem. However, as bandwidth demand increases, more and more progress towards this goal can be observed. In Japan, for instance EPON has largely replaced DSL as a broadband Internet source. South Korea's KT also provides a service called FTTH (Fibre To The Home), which provides fibre-optic connections to the subscriber's home. The largest FTTH deployments are in Japan, South Korea, and China. Singapore started implementation of their all-fibre Next Generation Nationwide Broadband Network (Next Gen NBN), which is slated for completion in 2012 and is being installed by OpenNet. Since they began rolling out services in September 2010, Network coverage in Singapore has reached 85 per cent nationwide.

In the US, Verizon Communications provides a FTTH service called FiOS to select high-ARPU (Average Revenue Per User) markets within its existing territory. The other major surviving ILEC (or Incumbent Local Exchange Carrier), AT and T, uses a FTTN (Fibre To The Node) service called U-verse with twisted-pair to the home. Their MSO competitors employ FTTN with coax using HFC. All of the major access networks use fibre for the bulk of the distance from the service provider's network to the customer.

Also in the US, Wilson Utilities located in Wilson, North Carolina, have implemented FTTH and have successfully achieved 1 gigabit fibre to the home. This was implemented in late 2013. Wilson Utilities first rolled out their FTTN (Fibre to the Home) in 2012 with speeds offerings of 20/40/60/100 megabits per second. Their service is referred to as GreenLight.

The globally dominant access network technology is EPON (Ethernet Passive Optical Network). In Europe, and among telcos in the United States, BPON (ATM-based Broadband PON) and GPON (Gigabit PON) had roots in the FSAN (Full Service Access Network) and ITU-T standards organizations under their control.

Comparison with electrical transmission

The choice between optical fibre and electrical (or copper) transmission for a particular system is made based on a number of trade-offs. Optical fibre is generally chosen for systems requiring higher bandwidth or spanning longer distances than electrical cabling can accommodate.

The main benefits of fibre are its exceptionally low loss (allowing long distances between amplifiers/repeaters), its absence of ground currents and other parasite signal and power issues common to long parallel electric conductor runs (due to its reliance on light rather than electricity for transmission, and the dielectric nature of fibre optic), and its inherently high data-carrying capacity. Thousands of electrical links would be required to replace a single high bandwidth fibre cable. Another benefit of fibers is that even when run alongside each other for long distances, fibre cables experience effectively no crosstalk, in contrast to some types of electrical transmission lines. Fibre can be installed in areas with high electromagnetic interference (EMI), such as alongside utility lines, power lines, and railroad tracks. Nonmetallic all-dielectric cables are also ideal for areas of high lightning-strike incidence.

For comparison, while single-line, voice-grade copper systems longer than a couple of kilometers require in-line signal repeaters for satisfactory performance; it is not unusual for optical systems to go over 100 kilometers

(62 mi), with no active or passive processing. Single-mode fibre cables are commonly available in 12 km lengths, minimizing the number of splices required over a long cable run. Multi-mode fibre is available in lengths up to 4 km, although industrial standards only mandate 2 km unbroken runs.

In short distance and relatively low bandwidth applications, electrical transmission is often preferred because of its

- Lower material cost, where large quantities are not required
- Lower cost of transmitters and receivers
- Capability to carry electrical power as well as signals (in appropriately designed cables)
- Ease of operating transducers in linear mode.
- Crosstalk from nearby cables and other parasitical unwanted signals increase profits from replacement and mitigation devices.

Optical fibers are more difficult and expensive to splice than electrical conductors. And at higher powers, optical fibers are susceptible to fibre fuse, resulting in catastrophic destruction of the fibre core and damage to transmission components.

Because of these benefits of electrical transmission, optical communication is not common in short box-to-box, backplane, or chip-to-chip applications; however, optical systems on those scales have been demonstrated in the laboratory.

In certain situations fibre may be used even for short distance or low bandwidth applications, due to other important features:

- Immunity to electromagnetic interference, including nuclear electromagnetic pulses (although fibre can be damaged by alpha and beta radiation).
- High electrical resistance, making it safe to use near high-voltage equipment or between areas with different earth potentials.
- Lighter weight—important, for example, in aircraft.
- No sparks—important in flammable or explosive gas environments.
- Not electromagnetically radiating, and difficult to tap without disrupting the signal—important in high-security environments.
- Much smaller cable size—important where pathway is limited, such as networking an existing building, where smaller channels can be drilled and space can be saved in existing cable ducts and trays.
- Resistance to corrosion due to non-metallic transmission medium

Optical fibre cables can be installed in buildings with the same equipment that is used to install copper and coaxial cables, with some modifications due to the small size and limited pull tension and bend radius of optical cables. Optical cables can typically be installed in duct systems in spans of 6000 meters or more depending on the duct's condition, layout of the duct system, and installation technique. Longer cables can be coiled at an intermediate point and pulled farther into the duct system as necessary.

Governing standards

In order for various manufacturers to be able to develop components that function compatibly in fibre optic communication systems, a number of standards have been developed. The International Telecommunications Union publishes several standards related to the characteristics and performance of fibers themselves, including

- ITU-T G.651, "Characteristics of a 50/125 μm multimode graded index optical fibre cable"
- ITU-T G.652, "Characteristics of a single-mode optical fibre cable"

Other standards specify performance criteria for fibre, transmitters, and receivers to be used together in conforming systems. Some of these standards are:

- 100 Gigabit Ethernet
- 10 Gigabit Ethernet
- Fibre Channel
- Gigabit Ethernet
- HIPPI
- Synchronous Digital Hierarchy
- Synchronous Optical Networking
- Optical Transport Network (OTN)

TOSLINK is the most common format for digital audio cable using plastic optical fibre to connect digital sources to digital receivers.

FIBRE OPTIC SENSOR

A fibre optic sensor is a sensor that uses optical fibre either as the sensing element ("intrinsic sensors"), or as a means of relaying signals from a remote sensor to the electronics that process the signals ("extrinsic sensors"). Fibers have many uses in remote sensing. Depending on the application, fibre may be used because of its small size, or because no electrical power is needed at the remote location, or because many sensors can be multiplexed along the length of a fibre by using light wavelength shift for each sensor, or by sensing the time delay as light passes along the fibre through each sensor. Time delay can be determined using a device such as an optical time-domain reflectometer and wavelength shift can be calculated using an instrument implementing optical frequency domain reflectometry.

Fibre optic sensors are also immune to electromagnetic interference, and do not conduct electricity so they can be used in places where there is high voltage electricity or flammable material such as jet fuel. Fibre optic sensors can be designed to withstand high temperatures as well.

Intrinsic sensors

Optical fibers can be used as sensors to measure strain,temperature, pressure and other quantities by modifying a fibre so that the quantity to be measured modulates the intensity, phase, polarization, wavelength or transit time of light in the fibre. Sensors that vary the intensity of light are the simplest, since only a simple source and detector are required. A particularly useful feature of intrinsic fibre optic sensors is that they can, if required, provide distributed sensing over very large distances.

Temperature can be measured by using a fibre that has evanescent loss that varies with temperature, or by analyzing the Raman scattering of the optical fibre. Electrical voltage can be sensed by nonlinear optical effects in specially-doped fibre, which alter the polarization of light as a function of voltage or electric field. Angle measurement sensors can be based on the Sagnac effect.

Special fibers like long-period fibre grating (LPG) optical fibers can be used for direction recognition . Photonics Research Group of Aston University in UK has some publications on vectorial bend sensor applications.

Optical fibers are used as hydrophones for seismic and sonar applications. Hydrophone systems with more than one hundred sensors per fibre cable have been developed. Hydrophone sensor systems are used by the oil industry as well as a few countries' navies. Both bottom-mounted hydrophone arrays and towed streamer systems are in use. The German company Sennheiser developed a laser microphone for use with optical fibers.

A fibre optic microphone and fibre-optic based headphone are useful in areas with strong electrical or magnetic fields, such as communication amongst the team of people working on a patient inside a magnetic resonance imaging (MRI) machine during MRI-guided surgery.

Optical fibre sensors for temperature and pressure have been developed for downhole measurement in oil wells. The fibre optic sensor is well suited for this environment as it functions at temperatures too high for semiconductor sensors (distributed temperature sensing).

Optical fibers can be made into interferometric sensors such as fibre optic gyroscopes, which are used in the Boeing 767 and in some car models (for navigation purposes). They are also used to make hydrogen sensors.

Fibre-optic sensors have been developed to measure co-located temperature and strain simultaneously with very high accuracy using fibre Bragg gratings. This is particularly useful when acquiring information from small complex structures. Brillouin scattering effects can be used to detect strain and temperature over larger distances (20–30 kilometers).

Other examples

A fibre-optic AC/DC voltage sensor in the middle and high voltage range (100–2000 V) can be created by inducing measurable amounts of Kerr nonlinearity in single mode optical fibre by exposing a calculated length of fibre to the external electric field. The measurement technique is based on polarimetric detection and high accuracy is achieved in a hostile industrial environment. High frequency (5 MHz–1 GHz) electromagnetic fields can be detected by induced nonlinear effects in fibre with a suitable structure. The fibre used is designed such that the Faraday and Kerr effects cause considerable phase change in the presence of the external field. With appropriate sensor design, this type of fibre can be used to measure different electrical and magnetic quantities and different internal parameters of fibre material. Electrical power can be measured in a fibre by using a structured bulk fibre ampere sensor coupled with proper signal processing in a polarimetric detection scheme. Experiments have been carried out in support of the technique. Fibre-optic sensors are used in electrical switchgear to transmit light from an electrical arc flash to a digital protective relay to enable fast tripping of a breaker to reduce the energy in the arc blast.

Extrinsic sensors

Extrinsic fibre optic sensors use an optical fibre cable, normally a multimode one, to transmit modulated light from either a non-fibre optical sensor, or an electronic sensor connected to an optical transmitter. A major benefit of extrinsic sensors is their ability to reach places which are otherwise inaccessible. An example is the measurement of temperature inside aircraft jet engines by using a fibre to transmit radiation into a radiation pyrometer located outside the engine. Extrinsic sensors can also be used in the same way to measure the internal temperature of electrical transformers, where the extreme electromagnetic fields present make other measurement techniques impossible.

Extrinsic fibre optic sensors provide excellent protection of measurement signals against noise corruption. Unfortunately, many conventional sensors produce electrical output which must be converted into an optical signal for use with fibre. For example, in the case of a platinum resistance thermometer, the temperature changes are translated into resistance changes. The PRT must therefore have an electrical power supply. The modulated voltage level at the output of the PRT can then be injected into the optical fibre via the usual type of transmitter. This complicates the measurement process and means that low-voltage power cables must be routed to the transducer. Extrinsic sensors are used to measure vibration, rotation, displacement, velocity, acceleration, torque, and temperature.

OTHER USES

Optical fibers have a wide number of applications. They are used as light guides in medical and other applications where bright light needs to be shone on a target without a clear line-of-sight path. In some buildings, optical fibers route sunlight from the roof to other parts of the building (see nonimaging optics). Optical fibre lamps are used for illumination in decorative applications, including signs, art, toys and artificial Christmas trees. Swarovski boutiques use optical fibers to illuminate their crystal showcases from many different angles while only employing one light source. Optical fibre is an intrinsic part of the light-transmitting concrete building product, LiTraCon.

Optical fibre is also used in imaging optics. A coherent bundle of fibers is used, sometimes along with lenses, for a long, thin imaging device called an endoscope, which is used to view objects through a small hole. Medical endoscopes are used for minimally invasive exploratory or surgical procedures. Industrial endoscopes (see fiberscope or borescope) are used for inspecting anything hard to reach, such as jet engine interiors. Many microscopes use fibre-optic light sources to provide intense illumination of samples being studied.

In spectroscopy, optical fibre bundles transmit light from a spectrometer to a substance that cannot be placed inside the spectrometer itself, in order to analyze its composition. A spectrometer analyzes substances by bouncing light off and through them. By using fibers, a spectrometer can be used to study objects remotely.

An optical fibre doped with certain rare earth elements such as erbium can be used as the gain medium of a laser or optical amplifier. Rare-earth-doped optical fibers can be used to provide signal amplification by splicing a short section of doped fibre into a regular (undoped) optical fibre line. The doped fibre is optically pumped with a second laser wavelength that is coupled into the line in addition to the signal wave. Both wavelengths of light are transmitted through the doped fibre, which transfers energy from the second pump wavelength to the signal wave. The process that causes the amplification is stimulated emission.

Optical fibre is also widely exploited as a nonlinear medium. The glass medium supports a host of nonlinear optical interactions, and the long interaction lengths possible in fibre facilitate a variety of phenomena, which are harnessed for applications and fundamental investigation. Conversely, fibre nonlinearity can have deleterious effects on optical signals, and measures are often required to minimize such unwanted effects. Optical fibers doped with a wavelength shifter collect scintillation light in physics experiments. Fibre optic sights for handguns, rifles, and shotguns use pieces of optical fibre to improve visibility of markings on the sight.

PRINCIPLE OF OPERATION

An optical fibre is a cylindrical dielectric waveguide (nonconducting waveguide) that transmits light along its axis, by the process of total internal reflection. The fibre consists of a *core* surrounded by a cladding layer, both of which are made of dielectric materials. To confine the optical signal in the core, the refractive index of the core must be greater than that of the cladding. The boundary between the core and cladding may either be abrupt, in *step-index fibre*, or gradual, in *graded-index fibre*.

REFRACTIVE INDEX

In optics the refractive index or index of refraction n of an optical medium is a dimensionless number that describes how light, or any other radiation, propagates through that medium. It is defined as

$$n = \frac{c}{v},$$

where c is the speed of light in vacuum and v is the speed of light in the substance. For example, the refractive index of water is 1.33, meaning that light travels 1.33 times faster in a vacuum than it does in water.

The refractive index determines how much light is bent, or refracted, when entering a material. This is the historically first use of refractive indices and is described by Snell's law of refraction, $n_1 \sin\theta_1 = n_2 \sin\theta_2$, where θ_1 and θ_2 are the angles of incidence and refraction, respectively, of a ray crossing the interface between two media with refractive indices n_1 and n_2. The refractive indices also determine the amount of light that is reflected when reaching the interface, as well as the critical angle for total internal reflection and Brewster's angle.

The refractive index can be seen as the factor by which the speed and the wavelength of the radiation are reduced with respect to their vacuum values: the speed of light in a medium is $v = c / n$, and similarly the wavelength in that medium is $\lambda = \lambda_0 / n$, where λ_0 is the wavelength of that light in vacuum. This implies that vacuum has a refractive index of 1, and that the frequency $(f = v / \lambda)$ of the wave is not affected by the refractive index.

The refractive index varies with the wavelength of light. This is called dispersion and causes the splitting of white light into its constituent colors in prisms and rainbows, and chromatic aberration in lenses. Light propagation in absorbing materials can be described using a complex-valued refractive index. The imaginary part then handles the attenuation, while the real part accounts for refraction.

The concept of refractive index is widely used within the full electromagnetic spectrum, from x-rays to radio waves. It can also be used with wave phenomena such as sound. In this case the speed of sound is used instead of that of light and a reference medium other than vacuum must be chosen.

Definition

The refractive index n of an optical medium is defined as the ratio of the speed of light in vacuum, c = 299792458 m/s, and the phase velocity v_{phase} of light in the medium,

$$n = \frac{c}{v_{\text{phase}}}.$$

The phase velocity is the speed at which the crests and the phase of the wave moves, which may be different from the group velocity, the speed at which the pulse of light, or the envelope of the wave, moves.

The definition above is sometimes referred to as the absolute index of refraction to distinguish it from definitions where the speed of light in other reference media than vacuum is used. Historically air at a standardized pressure and temperature have been common as a reference medium.

History

Thomas Young was presumably the person who first used, and invented, the name "index of refraction", in 1807. At the same time he changed this value of refractive power into a single number, instead of the traditional ratio of two numbers. The ratio had the disadvantage of different appearances. Newton, who called it the "proportion of the sines of incidence and refraction", wrote it as a ratio of two numbers, like "529 to 396" (or "nearly 4 to 3"; for water). Hauksbee, who called it the "ratio of refraction", wrote it as a ratio with a fixed numerator, like "10000 to 7451.9" (for urine). Hutton wrote it as a ratio with a fixed denominator, like 1.3358 to 1 (water). Young did not use a symbol for the index of refraction, in 1807. In the next years, others started using different symbols: n, m, and μ. The symbol n gradually prevailed.

Typical values

For visible light most transparent media have refractive indices between 1 and 2. These values are measured at the yellow doublet sodium D-line, with a wavelength of 589 nanometers, as is conventionally done. Gases at atmospheric pressure have refractive indices close to 1 because of their low density. Almost all solids and liquids have refractive indices above 1.3, with aerogel as the clear exception. Aerogel is a very low density solid that can be produced with refractive index in the range from 1.002 to 1.265. Diamond lies at the other end of the range with a refractive index as high as 2.42. Most plastics have refractive indices in the range from 1.3 to 1.7, but some high-refractive-index polymers can have values as high as 1.76.

For infrared light refractive indices can be considerably higher. Germanium is transparent in the wavelength region from 2 to 14 μm and has a refractive index of about 4, making it an important material for infrared optics.

Refractive index below 1

A widespread misconception is that since, according to the theory of relativity, nothing can travel faster than the speed of light in vacuum, the refractive index cannot be lower than 1. This is erroneous since the refractive index measures the phase velocity of light, which does not carry information. The phase velocity is the speed at which the crests of the wave move and can be faster than the speed of light in vacuum, and thereby give a refractive index below 1. This can occur close to resonance frequencies, for absorbing media, in plasmas, and for x-rays. In the x-ray regime the refractive indices are lower than but very close to 1 (exceptions close to some resonance frequencies). As an example, water has a refractive index of 0.99999974 = 1 " 2.6×10 for x-ray radiation at a photon energy of 30 keV (0.04 nm wavelength).

Negative refractive index

Recent research has also demonstrated the existence of materials with a negative refractive index, which can occur if permittivity and permeability have simultaneous negative values. This can be achieved with periodically constructed metamaterials. The resulting negative refraction (*i.e.*, a reversal of Snell's law) offers the possibility of the superlens and other exotic phenomena.

Microscopic explanation

At the microscale, an electromagnetic wave's phase velocity is slowed in a material because the electric field creates a disturbance in the charges of each atom (primarily the electrons) proportional to the electric susceptibility of the medium. (Similarly, the magnetic field creates a disturbance proportional to the magnetic susceptibility.) As the electromagnetic fields oscillate in the wave, the charges in the material will be "shaken" back and forth at the same frequency. The charges thus radiate their own electromagnetic wave that is at the same frequency, but usually with a phase delay, as the charges may move out of phase with the force driving them (see sinusoidally driven harmonic oscillator). The light wave traveling in the medium is the macroscopic superposition (sum) of all such contributions in the material: the original wave plus the waves radiated by all the moving charges. This wave is typically a wave with the same frequency but shorter wavelength than the original, leading to a slowing of the wave's phase velocity. Most of the radiation from oscillating material charges will modify the incoming wave, changing its velocity. However, some net energy will be radiated in other directions or even at other frequencies (see scattering).

Depending on the relative phase of the original driving wave and the waves radiated by the charge motion, there are several possibilities:

- If the electrons emit a light wave which is 90° out of phase with the light wave shaking them, it will cause the total light wave to travel more slowly. This is the normal refraction of transparent materials like glass or water, and corresponds to a refractive index which is real and greater than 1.
- If the electrons emit a light wave which is 270° out of phase with the light wave shaking them, it will cause the total light wave to travel

more quickly. This is called "anomalous refraction", and is observed close to absorption lines, with x-rays, and in some microwave systems. It corresponds to a refractive index less than 1. (Even though the phase velocity of light is greater than the speed of light in vacuum *c*, the signal velocity is not, as discussed above). If the response is sufficiently strong and out-of-phase, the result is a negative refractive index.

- If the electrons emit a light wave which is 180° out of phase with the light wave shaking them, it will destructively interfere with the original light to reduce the total light intensity. This is light absorption in opaque materials and corresponds to an imaginary refractive index.
- If the electrons emit a light wave which is in phase with the light wave shaking them, it will amplify the light wave. This is rare, but occurs in lasers due to stimulated emission. It corresponds to an imaginary index of refraction, with the opposite sign to that of absorption.

For most materials at visible-light frequencies, the phase is somewhere between 90° and 180°, corresponding to a combination of both refraction and absorption.

Dispersion

The refractive index of materials varies with the wavelength (and frequency) of light. This is called dispersion and causes prisms and rainbows to divide white light into its constituent spectral colors. As the refractive index varies with wavelength, so will the refraction angle as light goes from one material to another. Dispersion also causes the focal length of lenses to be wavelength dependent. This is a type of chromatic aberration, which often needs to be corrected for in imaging systems. In regions of the spectrum where the material does not absorb light, the refractive index tends to decrease with increasing wavelength, and thus increase with frequency. This is called normal dispersion, in contrast to anomalous dispersion, where the refractive index increases with wavelength. For visible light normal dispersion means that the refractive index is higher for blue light than for red.

For optics in the visual range the amount of dispersion of a lens material is often quantified by the Abbe number $V = \frac{n_{\text{yellow}} - 1}{n_{\text{blue}} - n_{\text{red}}}$. For a more accurate description of the wavelength dependence of the refractive index the Sellmeier equation can be used.

Because of dispersion, it is usually important to specify the vacuum wavelength of light for which a refractive index is measured. Typically, measurements are done at various well-defined spectral emission lines; for example, n_D usually denotes the refractive index at the Fraunhofer "D" line, the centre of the yellow sodium double emission at 589.29 nm wavelength.

Complex index of refraction and absorption

When light passes through a medium, some part of it will always be absorbed. This can be conveniently taken into account by defining a complex index of refraction,

$$\tilde{n} = n + ik$$

Here, the real part of the refractive index n indicates the phase velocity, while the imaginary part k indicates the amount of absorption loss when the electromagnetic wave propagates through the material.

That k corresponds to absorption can be seen by inserting this refractive index into the expression for electric field of a plane electromagnetic wave traveling in the z-direction. We can do this by relating the wave number to the refractive index through $k = \frac{2\pi n}{\lambda_0}$, with λ_0 being the vacuum wavelength. With complex wave number $\tilde{k}$ and refractive index $n + ik$ this can be inserted into the plane wave expression as

$$E(z,t) = \mathrm{Re}(E_0 e^{i(\tilde{k}z-\omega t)}) = \mathrm{Re}(E_0 e^{i(2\pi(n+ik)z/\lambda_0-\omega t)}) = e^{-2\pi kz/\lambda_0}\,\mathrm{Re}(E_0 e^{i(kz-\omega t)})$$

Here we see that K gives an exponential decay, as expected from the Beer–Lambert law. Since intensity is proportional to the square of the electric field, the absorption coefficient becomes $\frac{4\pi k}{\lambda_0}$.

$\hat{e}$ is often called the extinction coefficient in physics although this has a different definition within chemistry. Both n and $\hat{e}$ are dependent on the frequency. In most circumstances $k > 0$ (light is absorbed) or $k = 0$ (light travels forever without loss). In special situations, especially in the gain medium of lasers, it is also possible that $k > 0$, corresponding to an amplification of the light.

An alternative convention uses $\tilde{n} = n - ik$ instead of $\tilde{n} = n + ik$, but where $k > 0$ still corresponds to loss. Therefore these two conventions are inconsistent and should not be confused. The difference is related to defining sinusoidal time dependence as $\mathrm{Re}(e^{-i\omega t})$ versus $\mathrm{Re}(e^{+i\omega t})$.

Dielectric loss and non-zero DC conductivity in materials cause absorption. Good dielectric materials such as glass have extremely low DC conductivity, and at low frequencies the dielectric loss is also negligible, resulting in almost no absorption $(k \approx 0)$. However, at higher frequencies (such as visible light), dielectric loss may increase absorption significantly, reducing the material's transparency to these frequencies.

The real, n, and imaginary, k, parts of the complex refractive index are related through the Kramers–Kronig relations. In 1986 A.R. Forouhi and I.

Bloomer deduced an equation describing k as a function of photon energy, E, applicable to amorphous materials. Forouhi and Bloomer then applied the Kramers-Kronig relation to derive the corresponding equation for n as a function of E. The same formalism was applied to crystalline materials by Forouhi and Bloomer in 1988.

The refractive index and extinction coefficient, n and k, cannot be measured directly.

They must be determined indirectly from measurable quantities that depend on them, such as reflectance, R, or transmittance, T, or ellipsometric parameters, ψ and δ. The determination of n and k from such measured quantities will involve developing a theoretical expression for R or T, or ψ and δ in terms of a valid physical model for n and k. By fitting the theoretical model to the measured R or T, or ø and ä using regression analysis, n and k can be deduced.

For x-ray and extreme ultraviolet radiation the complex refractive index deviates only slightly from unity and usually has a real part smaller than 1. It is therefore normally written as $\tilde{n} = 1 - \delta + i\beta$ (or $\tilde{n} = 1 - \delta - i\beta$).

Relations to other quantities

Optical path length

Optical path length (OPL) is the product of the geometric length d of the path light follows through a system, and the index of refraction of the medium through which it propagates, $\text{OPL} = nd$. This is an important concept in optics because it determines the phase of the light and governs interference and diffraction of light as it propagates. According to Fermat's principle, light rays can be characterized as those curves that optimize the optical path length.

Refraction

When light moves from one medium to another as in the figure to the right, it changes direction, *i.e.* it is refracted. If it goes from a medium with refractive index n_1 to one with refractive index n_2, with an incidence angle to the surface normal of θ_1, the refraction angle θ_2 can be calculated from Snell's law:

$$n_1 \sin\theta = n_2 \sin\theta_2 .$$

When light enters a material with higher refractive index, the angle of refraction will be smaller than the angle of incidence and the light will be refracted towards the normal of the surface. The higher the refractive index, the closer to the normal direction the light will travel. When passing into a medium with lower refractive index, the light will instead be refracted away from the normal, towards the surface.

Total internal reflection

If there is no angle θ_2 fulfilling Snell's law, *i.e.*,

$$\frac{n_1}{n_2}\sin\theta_1 > 1,$$

the light cannot be transmitted and will instead undergo total internal reflection. This occurs only when going to a less optically dense material, *i.e.*, one with lower refractive index. To get total internal reflection the angles of incidence θ_1 must be larger than the critical angle

$$\theta_c = \arcsin\left(\frac{n_2}{n_1}\right)$$

Reflectivity

Apart from the transmitted light there is also a reflected part. The reflection angle is equal to the incidence angle, and the amount of light that is reflected is determined by the reflectivity of the surface. The reflectivity can be calculated from the refractive index and the incidence angle with the Fresnel equations, which for normal incidence reduces to

$$R = \left|\frac{n_1 - n_2}{n_1 + n_2}\right|^2.$$

For common glass in air, $n_1 = 1$ and $n_2 = 1.5$, and thus about 4 per cent of the incident power is reflected. At other incidence angles the reflectivity will also depend on the polarization of the incoming light. At a certain angle called Brewster's angle, p-polarized light (light with the electric field in the plane of incidence) will be totally transmitted. Brewster's angle can be calculated from the two refractive indices of the interface as

$$\theta_B = \arctan\left(\frac{n_2}{n_1}\right).$$

Lenses

The focal length of a lens is determined by its refractive index n and the radii of curvature R_1 and R_2 of its surfaces. The power of a thin lens in air is given by the Lensmaker's formula:

$$\frac{1}{f} = (n-1)\left(\frac{1}{R_1} - \frac{1}{R_2}\right)$$

where f is the focal length of the lens.

Dielectric constant

The refractive index of electromagnetic radiation equals

$$n = \sqrt{\epsilon_r \mu_r} \,,$$

where ϵ_r is the material's relative permittivity, and $\grave{\imath}_r$ is its relative permeability. For most naturally occurring materials, $\grave{\imath}_r$ is very close to 1 at optical frequencies, therefore n is approximately $\sqrt{\epsilon_r}$.

The frequency dependent dielectric constant is simply the square of the (complex) refractive index in a non-magnetic medium (one with a relative permeability of unity). The refractive index is used for optics in Fresnel equations and Snell's law; while the dielectric constant is used in Maxwell's equations and electronics.

Where $\tilde{\epsilon}$ is the complex dielectric constant with real and imaginary parts ϵ_1 and ϵ_2, and n and k are the real and imaginary parts of the refractive index, all functions of frequency:

$$\tilde{\epsilon} = \epsilon_1 + i\,\epsilon_2 = (n + ik)^2$$

Conversion between refractive index and dielectric constant is done by:

$$\epsilon_1 = n^2 - k^2$$

$$\epsilon_2 = 2nk$$

$$n = \sqrt{\frac{\sqrt{\epsilon_1^2 + \epsilon_2^2} + \epsilon_1}{2}}$$

$$n = \sqrt{\frac{\sqrt{\epsilon_1^2 + \epsilon_2^2} + \epsilon_1}{2}}$$

Density

In general, the refractive index of a glass increases with its density. However, there does not exist an overall linear relation between the refractive index and the density for all silicate and borosilicate glasses. A relatively high refractive index and low density can be obtained with glasses containing light metal oxides such as Li_2O and MgO, while the opposite trend is observed with glasses containing PbO and BaO as seen in the diagram at the right.

Many oils (such as olive oil) and ethyl alcohol are examples of liquids which are more refractive, but less dense, than water, contrary to the general correlation between density and refractive index.

For gases, $n - 1$ is proportional to the density of the gas as long as the chemical composition does not change. This means that it is also proportional to the pressure and inversely proportional to the temperature for ideal gases.

Group index

Sometimes, a "group velocity refractive index", usually called the *group index* is defined:

$$n_g = \frac{c}{v_g}$$

where v_g is the group velocity. This value should not be confused with n, which is always defined with respect to the phase velocity. When the dispersion is small, the group velocity can be linked to the phase velocity by the relation

$$n_g = v - \lambda \frac{dv}{d\lambda}.$$

In this case the group index can thus be written in terms of the wavelength dependence of the refractive index as

$$n_g = \frac{n}{1 + \frac{\lambda}{n}\frac{dn}{d\lambda}},$$

where λ is the wavelength in the medium.

When the refractive index of a medium is known as a function of the vacuum wavelength (instead of the wavelength in the medium), the corresponding expressions for the group velocity and index are (for all values of dispersion)

$$v_g = c\left(n - \lambda_0 \frac{dn}{d\lambda_0}\right)^{-1}$$

$$n_g = n - \lambda_0 \frac{dn}{d\lambda_0}$$

where λ_0 is the wavelength in vacuum.

Momentum (Abraham–Minkowski controversy)

In 1908, Hermann Minkowski calculated the momentum of a refracted ray, p, where E is energy of the photon, c is the speed of light in vacuum and n is the refractive index of the medium as follows:

$$p = \frac{nE}{c}$$

In 1909, Max Abraham proposed the following formula for this calculation:

$$p = \frac{E}{nc}$$

A 2010 study suggested that *both* equations are correct, with the Abraham version being the kinetic momentum and the Minkowski version being the canonical momentum, and claims to explain the contradicting experimental results using this interpretation.

TOTAL INTERNAL REFLECTION

When light traveling in an optically dense medium hits a boundary at a steep angle (larger than the critical angle for the boundary), the light is completely reflected. This is called total internal reflection. This effect is used in optical fibers to confine light in the core. Light travels through the fibre core, bouncing back and forth off the boundary between the core and cladding. Because the light must strike the boundary with an angle greater than the critical angle, only light that enters the fibre within a certain range of angles can travel down the fibre without leaking out. This range of angles is called the acceptance cone of the fibre. The size of this acceptance cone is a function of the refractive index difference between the fiber's core and cladding.

In simpler terms, there is a maximum angle from the fibre axis at which light may enter the fibre so that it will propagate, or travel, in the core of the fibre. The sine of this maximum angle is the numerical aperture (NA) of the fibre. Fibre with a larger NA requires less precision to splice and work with than fibre with a smaller NA. Single-mode fibre has a small NA.

MULTI-MODE FIBRE

Multi-mode optical fibre is a type of optical fibre mostly used for communication over short distances, such as within a building or on a campus. Typical multimode links have data rates of 10 Mbit/s to 10 Gbit/s over link lengths of up to 600 meters (2000 feet) — more than sufficient for the majority of premises applications.

Applications

The equipment used for communications over multi-mode optical fibre is less expensive than that for single-mode optical fibre. Typical transmission speed and distance limits are 100 Mbit/s for distances up to 2 km (100BASE-FX), 1 Gbit/s up to 1000 m, and 10 Gbit/s up to 550 m.

Because of its high capacity and reliability, multi-mode optical fibre generally is used for backbone applications in buildings. An increasing number of users are taking the benefits of fibre closer to the user by running fibre to the desktop or to the zone. Standards-compliant architectures such as

Centralized Cabling and fibre to the telecom enclosure offer users the ability to leverage the distance capabilities of fibre by centralizing electronics in telecommunications rooms, rather than having active electronics on each floor.

Comparison with single-mode fibre

The main difference between multi-mode and single-mode optical fibre is that the former has much larger core diameter, typically 50–100 micrometers; much larger than the wavelength of the light carried in it. Because of the large core and also the possibility of large numerical aperture, multi-mode fibre has higher "light-gathering" capacity than single-mode fibre. In practical terms, the larger core size simplifies connections and also allows the use of lower-cost electronics such as light-emitting diodes (LEDs) and vertical-cavity surface-emitting lasers (VCSELs) which operate at the 850 nm and 1300 nm wavelength (single-mode fibers used in telecommunications operate at 1310 or 1550 nm and require more expensive laser sources. Single mode fibers exist for nearly all visible wavelengths of light). However, compared to single-mode fibers, the multi-mode fibre bandwidth–distance product limit is lower. Because multi-mode fibre has a larger core-size than single-mode fibre, it supports more than one propagation mode; hence it is limited by modal dispersion, while single mode is not.

The LED light sources sometimes used with multi-mode fibre produce a range of wavelengths and these each propagate at different speeds. This chromatic dispersion is another limit to the useful length for multi-mode fibre optic cable. In contrast, the lasers used to drive single-mode fibers produce coherent light of a single wavelength. Due to the modal dispersion, multi-mode fibre has higher pulse spreading rates than single mode fibre, limiting multi-mode fiber's information transmission capacity.

Single-mode fibers are most often used in high-precision scientific research because the allowance of only one propagation mode of the light makes the light easier to focus properly.

Jacket colour is sometimes used to distinguish multi-mode cables from single-mode ones. The standard TIA-598C recommends, for civilian applications, the use of a yellow jacket for single-mode fibre, and orange or aqua for multi-mode fibre, depending on type. Some vendors use violet to distinguish higher performance OM4 communications fibre from other types.

Types

Multi-mode fibers are described by their core and cladding diameters. Thus, 62.5/125 μm multi-mode fibre has a core size of 62.5 micrometres (μm) and a cladding diameter of 125 μm. The transition between the core and cladding can be sharp, which is called a step-index profile, or a gradual transition, which is called a graded-index profile. The two types have different dispersion characteristics and thus different effective propagation distance. Multi-mode fibers may be constructed with either graded or step-index profile.

In addition, multi-mode fibers are described using a system of classification determined by the ISO 11801 standard — OM1, OM2, and OM3 — which is based on the modal bandwidth of the multi-mode fibre. OM4 (defined in TIA-492-AAAD) was finalized in August 2009, and was published by the end of 2009 by the TIA. OM4 cable will support 125m links at 40 and 100 Gbit/s. The letters "OM" stand for *optical multi-mode*.

For many years 62.5/125 μm (OM1) and conventional 50/125 μm multi-mode fibre (OM2) were widely deployed in premises applications. These fibers easily support applications ranging from Ethernet (10 Mbit/s) to gigabit Ethernet (1 Gbit/s) and, because of their relatively large core size, were ideal for use with LED transmitters. Newer deployments often use laser-optimized 50/125 μm multi-mode fibre (OM3). Fibers that meet this designation provide sufficient bandwidth to support 10 Gigabit Ethernet up to 300 meters. Optical fibre manufacturers have greatly refined their manufacturing process since that standard was issued and cables can be made that support 10 GbE up to 400 meters. Laser optimized multi-mode fibre (LOMMF) is designed for use with 850 nm VCSELs.

The migration to LOMMF/OM3 has occurred as users upgrade to higher speed networks. LEDs have a maximum modulation rate of 622 Mbit/s because they can not be turned on/off fast enough to support higher bandwidth applications. VCSELs are capable of modulation over 10 Gbit/s and are used in many high speed networks.

Cables can sometimes be distinguished by jacket colour: for 62.5/125 μm (OM1) and 50/125 μm (OM2), orange jackets are recommended, while Aqua is recommended for 50/125 μm "laser optimized" OM3 and OM4 fibre.

VCSEL power profiles, along with variations in fibre uniformity, can cause modal dispersion which is measured by differential modal delay (DMD). Modal dispersion is caused by the different speeds of the individual modes in a light pulse. The net effect causes the light pulse to spread over distance, introducing intersymbol interference. The greater the length, the greater the modal dispersion. To combat modal dispersion, LOMMF is manufactured in a way that eliminates variations in the fibre which could affect the speed that a light pulse can travel. The refractive index profile is enhanced for VCSEL transmission and to prevent pulse spreading. As a result the fibers maintain signal integrity over longer distances, thereby maximizing the bandwidth.

SINGLE-MODE OPTICAL FIBRE

In fibre-optic communication, a single-mode optical fibre (SMF) is an optical fibre designed to carry light only directly down the fibre - the transverse mode. Modes are the possible solutions of the Helmholtz equation for waves, which is obtained by combining Maxwell's equations and the boundary conditions. These modes define the way the wave travels through space, *i.e.* how the wave is distributed in space. Waves can have the same mode but have different frequencies. This is the case in single-mode fibers, where we can have waves with different frequencies, but of the same mode, which means that they are distributed in space in the same way, and that gives us a single ray of light. Although the ray travels parallel to the length of the fibre, it is often called transverse mode since its electromagnetic vibrations occur perpendicular (transverse) to the length of the fibre. The 2009 Nobel Prize in Physics was awarded to Charles K. Kao for his theoretical work on the single-mode optical fibre.

Professor Huang Hongjia of the Chinese Academy of Sciences, developed coupling wave theory in the field of microwave theory. He led a research team that successfully developed Single-mode optical fibre in 1980.

Characteristics

Like multi-mode optical fibers, single mode fibers do exhibit modal dispersion resulting from multiple spatial modes but with narrower modal dispersion. Single mode fibers are therefore better at retaining the fidelity of each light pulse over longer distances than multi-mode fibers. For these reasons, single-mode fibers can have a higher bandwidth than multi-mode fibers. Equipment for single mode fibre is more expensive than equipment for multi-mode optical fibre, but the single mode fibre itself is usually cheaper in bulk.

A typical single mode optical fibre has a core diameter between 8 and 10.5 μm and a cladding diameter of 125 μm. There are a number of special types of single-mode optical fibre which have been chemically or physically altered to give special properties, such as dispersion-shifted fibre and nonzero dispersion-shifted fibre. Data rates are limited by polarization mode dispersion and chromatic dispersion. As of 2005, data rates of up to 10 gigabits per second were possible at distances of over 80 km (50 mi) with commercially available transceivers (Xenpak). By using optical amplifiers and dispersion-compensating devices, state-of-the-art DWDM optical systems can span thousands of kilometers at 10 Gbit/s, and several hundred kilometers at 40 Gbit/s.

The lowest-order bounds mode is ascertained for the wavelength of interest by solving Maxwell's equations for the boundary conditions imposed by the fibre, which are determined by the core diameter and the refractive indices of the core and cladding. The solution of Maxwell's equations for the lowest order bound mode will permit a pair of orthogonally polarized fields in the fibre, and this is the usual case in a communication fibre.

In step-index guides, single-mode operation occurs when the normalized frequency, V, is less than or equal to 2.405. For power-law profiles, single-mode operation occurs for a normalized frequency, V, less than approximately

$$2.405\sqrt{\frac{g+2}{g}},$$

where g is the profile parameter.

In practice, the orthogonal polarizations may not be associated with degenerate modes.

OS1 and OS2 are standard single-mode optical fibre used with wavelengths 1310 nm and 1550 nm (size 9/125 μm) with a maximum attenuation of 1 dB/km (OS1) and .4 dB/km (OS2). OS1 is defined in ISO/IEC 11801, and OS2 is defined in ISO/IEC 24702.

Connectors

Optical fibre connectors are used to join optical fibers where a connect/disconnect capability is required. The basic connector unit is a connector assembly. A connector assembly consists of an adapter and two connector plugs. Due to the sophisticated polishing and tuning procedures that may be incorporated into optical connector manufacturing, connectors are generally assembled onto optical fibre in a supplier's manufacturing facility. However, the assembly and polishing operations involved can be performed in the field, for example to make cross-connect jumpers to size.

Optical fibre connectors are used in telephone company central offices, at installations on customer premises, and in outside plant applications. Their uses include:

- Making the connection between equipment and the telephone plant in the central office
- Connecting fibers to remote and outside plant electronics such as Optical Network Units (ONUs) and Digital Loop Carrier (DLC) systems
- Optical cross connects in the central office
- Patching panels in the outside plant to provide architectural flexibility and to interconnect fibers belonging to different service providers
- Connecting couplers, splitters, and Wavelength Division Multiplexers (WDMs) to optical fibers
- Connecting optical test equipment to fibers for testing and maintenance.

Outside plant applications may involve locating connectors underground in subsurface enclosures that may be subject to flooding, on outdoor walls, or on utility poles. The closures that enclose them may be hermetic, or may be "free-breathing." Hermetic closures will prevent subjection of the connectors within to temperature swings unless they are breached. Free-breathing enclosures will subject them to temperature and humidity swings, and possibly to condensation and biological action from airborne bacteria, insects, etc.

Connectors in the underground plant may be subjected to groundwater immersion if the closures containing them are breached or improperly assembled.

The latest industry requirements for optical fibre connectors are in Telcordia GR-326, *Generic Requirements for Singlemode Optical Connectors and Jumper Assemblies*.

A *multi-fibre* optical connector is designed to simultaneously join multiple optical fibers together, with each optical fibre being joined to only one other optical fibre.

The last part of the definition is included so as not to confuse multi-fibre connectors with a branching component, such as a coupler. The latter joins one optical fibre to two or more other optical fibers.

Multi-fibre optical connectors are designed to be used wherever quick and/ or repetitive connects and disconnects of a group of fibers are needed. Applications include telecommunications companies' Central Offices (COs), installations on customer premises, and Outside Plant (OSP) applications.

The multi-fibre optical connector can be used in the creation of a low-cost switch for use in fibre optical testing. Another application is in cables delivered to a user with pre-terminated multi-fibre jumpers. This would reduce the need for field splicing, which could greatly reduce the amount of hours necessary for placing an optical fibre cable in a telecommunications network. This, in turn, would result in savings for the installer of such cable. Industry requirements for multi-fibre optical connectors are covered in GR-1435, *Generic Requirements for Multi-Fibre Optical Connectors*.

Fibre Optic Switches

An optical switch is a component with two or more ports that selectively transmits, redirects, or blocks an optical signal in a transmission medium. According to Telcordia GR-1073, an optical switch must be actuated to select or change between states. The actuating signal (also referred to as the control signal) is usually electrical, but in principle, could be optical or mechanical. (The control signal format may be Boolean and may be a separate signal; or, in the case of optical actuation, the control signal may be encoded in the input data signal. Switch performance is generally intended to be independent of wavelength within the component passband.)

SPECIAL-PURPOSE FIBRE

Some special-purpose optical fibre is constructed with a non-cylindrical core and/or cladding layer, usually with an elliptical or rectangular cross-section. These include polarization-maintaining fibre and fibre designed to suppress whispering gallery mode propagation. Polarization-maintaining fibre is a unique type of fibre that is commonly used in fibre optic sensors due to its ability to maintain the polarization of the light inserted into it.

Photonic-crystal fibre is made with a regular pattern of index variation (often in the form of cylindrical holes that run along the length of the fibre). Such fibre uses diffraction effects instead of or in addition to total internal reflection, to confine light to the fiber's core. The properties of the fibre can be tailored to a wide variety of applications.

MECHANISMS OF ATTENUATION

Attenuation in fibre optics, also known as transmission loss, is the reduction in intensity of the light beam (or signal) as it travels through the transmission medium. Attenuation coefficients in fibre optics usually use units of dB/km through the medium due to the relatively high quality of transparency of modern optical transmission media. The medium is usually a fibre of silica glass that confines the incident light beam to the inside. Attenuation is an important factor limiting the transmission of a digital signal across large distances. Thus, much research has gone into both limiting the attenuation and maximizing the amplification of the optical signal. Empirical research has shown that attenuation in optical fibre is caused primarily by both scattering and absorption. Single-mode optical fibers can be made with extremely low loss. Corning's SMF-28 fibre, a standard single-mode fibre for telecommunications wavelengths, has a loss of 0.17 dB/km at 1550 nm. For example, an 8 km length of SMF-28 transmits nearly 75 per cent of light at 1550 nm. It has been noted that if ocean water was as clear as fibre, one could see all the way to the bottom even of the Marianas Trench in the Pacific Ocean, a depth of 36,000 feet.

LIGHT SCATTERING

The propagation of light through the core of an optical fibre is based on total internal reflection of the lightwave. Rough and irregular surfaces, even at the molecular level, can cause light rays to be reflected in random directions. This is called diffuse reflection or scattering, and it is typically characterized by wide variety of reflection angles.

Light scattering depends on the wavelength of the light being scattered. Thus, limits to spatial scales of visibility arise, depending on the frequency of the incident light-wave and the physical dimension (or spatial scale) of the scattering center, which is typically in the form of some specific micro-structural feature. Since visible light has a wavelength of the order of one micrometer (one millionth of a meter) scattering centers will have dimensions on a similar spatial scale.

Thus, attenuation results from the incoherent scattering of light at internal surfaces and interfaces. In (poly)crystalline materials such as metals and ceramics, in addition to pores, most of the internal surfaces or interfaces are in the form of grain boundaries that separate tiny regions of crystalline order. It has recently been shown that when the size of the scattering center (or grain boundary) is reduced below the size of the wavelength of the light being scattered, the scattering no longer occurs to any significant extent. This phenomenon has given rise to the production of transparent ceramic materials.

Similarly, the scattering of light in optical quality glass fibre is caused by molecular level irregularities (compositional fluctuations) in the glass structure. Indeed, one emerging school of thought is that a glass is simply the limiting case of a polycrystalline solid. Within this framework, "domains" exhibiting various degrees of short-range order become the building blocks of both metals and alloys, as well as glasses and ceramics. Distributed both between and within these domains are micro-structural defects that provide the most ideal locations for light scattering. This same phenomenon is seen as one of the limiting factors in the transparency of IR missile domes. At high optical powers, scattering can also be caused by nonlinear optical processes in the fibre.

UV-VIS-IR ABSORPTION

In addition to light scattering, attenuation or signal loss can also occur due to selective absorption of specific wavelengths, in a manner similar to that responsible for the appearance of colour. Primary material considerations include both electrons and molecules as follows:

- At the electronic level, it depends on whether the electron orbitals are spaced (or "quantized") such that they can absorb a quantum of light (or photon) of a specific wavelength or frequency in the ultraviolet (UV) or visible ranges. This is what gives rise to colour.
- At the atomic or molecular level, it depends on the frequencies of atomic or molecular vibrations or chemical bonds, how close-packed its atoms or molecules are, and whether or not the atoms or molecules exhibit long-range order. These factors will determine the capacity of the material transmitting longer wavelengths in the infrared (IR), far IR, radio and microwave ranges.

The design of any optically transparent device requires the selection of materials based upon knowledge of its properties and limitations. The Lattice absorption characteristics observed at the lower frequency regions (mid IR to far-infrared wavelength range) define the long-wavelength transparency limit of the material. They are the result of the interactive coupling between the motions of thermally induced vibrations of the constituent atoms and molecules of the solid lattice and the incident light wave radiation. Hence, all materials are bounded by limiting regions of absorption caused by atomic and molecular vibrations (bond-stretching)in the far-infrared ($>10\ \mu m$).

Thus, multi-phonon absorption occurs when two or more phonons simultaneously interact to produce electric dipole moments with which the incident radiation may couple. These dipoles can absorb energy from the incident radiation, reaching a maximum coupling with the radiation when the frequency is equal to the fundamental vibrational mode of the molecular dipole (*e.g.*. Si-O bond) in the far-infrared, or one of its harmonics.

The selective absorption of infrared (IR) light by a particular material occurs because the selected frequency of the light wave matches the frequency (or an integer multiple of the frequency) at which the particles of that material vibrate. Since different atoms and molecules have different natural frequencies of vibration, they will selectively absorb different frequencies (or portions of the spectrum) of infrared (IR) light.

Reflection and transmission of light waves occur because the frequencies of the light waves do not match the natural resonant frequencies of vibration of the objects. When IR light of these frequencies strikes an object, the energy is either reflected or transmitted.

MANUFACTURING

MATERIALS

Glass optical fibers are almost always made from silica, but some other materials, such as fluorozirconate, fluoroaluminate, and chalcogenide glasses as well as crystalline materials like sapphire, are used for longer-wavelength infrared or other specialized applications. Silica and fluoride glasses usually have refractive indices of about 1.5, but some materials such as the chalcogenides can have indices as high as 3. Typically the index difference between core and cladding is less than one percent.

Plastic optical fibers (POF) are commonly step-index multi-mode fibers with a core diameter of 0.5 millimeters or larger. POF typically have higher attenuation coefficients than glass fibers, 1 dB/m or higher, and this high attenuation limits the range of POF-based systems.

SILICA

Silica exhibits fairly good optical transmission over a wide range of wavelengths. In the near-infrared (near IR) portion of the spectrum, particularly around 1.5 ìm, silica can have extremely low absorption and scattering losses of the order of 0.2 dB/km. Such remarkably low losses are possible only because ultra-pure silicon is available, it being essential for manufacturing integrated circuits and discrete transistors. A high transparency in the 1.4-ìm region is achieved by maintaining a low concentration of hydroxyl groups (OH). Alternatively, a high OH concentration is better for transmission in the ultraviolet (UV) region.

Silica can be drawn into fibers at reasonably high temperatures, and has a fairly broad glass transformation range. One other advantage is that fusion splicing and cleaving of silica fibers is relatively effective. Silica fibre also has high mechanical strength against both pulling and even bending, provided that the fibre is not too thick and that the surfaces have been well prepared during processing. Even simple cleaving (breaking) of the ends of the fibre can provide nicely flat surfaces with acceptable optical quality. Silica is also relatively chemically inert. In particular, it is not hygroscopic (does not absorb water).

Silica glass can be doped with various materials. One purpose of doping is to raise the refractive index (*e.g.*. with germanium dioxide (GeO_2) or aluminium oxide (Al_2O_3)) or to lower it (*e.g.*. with fluorine or boron trioxide (B_2O_3)). Doping is also possible with laser-active ions (for example, rare earth-doped fibers) in order to obtain active fibers to be used, for example, in fibre amplifiers or laser applications. Both the fibre core and cladding are typically doped, so that the entire assembly (core and cladding) is effectively the same compound (*e.g.*. an aluminosilicate, germanosilicate, phosphosilicate or borosilicate glass).

Particularly for active fibers, pure silica is usually not a very suitable host glass, because it exhibits a low solubility for rare earth ions. This can lead to quenching effects due to clustering of dopant ions. Aluminosilicates are much more effective in this respect.

Silica fibre also exhibits a high threshold for optical damage. This property ensures a low tendency for laser-induced breakdown. This is important for fibre amplifiers when utilized for the amplification of short pulses.

Because of these properties silica fibers are the material of choice in many optical applications, such as communications (except for very short distances with plastic optical fibre), fibre lasers, fibre amplifiers, and fibre-optic sensors. Large efforts put forth in the development of various types of silica fibers have further increased the performance of such fibers over other materials.

FLUORIDE GLASS

Fluoride glass is a class of non-oxide optical quality glasses composed of fluorides of various metals. Because of their low viscosity, it is very difficult to completely avoid crystallization while processing it through the glass transition (or drawing the fibre from the melt). Thus, although heavy metal fluoride glasses (HMFG) exhibit very low optical attenuation, they are not only difficult to manufacture, but are quite fragile, and have poor resistance to moisture and other environmental attacks. Their best attribute is that they lack the absorption band associated with the hydroxyl (OH) group (3200–3600 cm; *i.e.*, 2777–3125 nm or 2.78–3.13 μm), which is present in nearly all oxide-based glasses.

An example of a heavy metal fluoride glass is the ZBLAN glass group, composed of zirconium, barium, lanthanum, aluminium, and sodium fluorides. Their main technological application is as optical waveguides in both planar and fibre form. They are advantageous especially in the mid-infrared (2000–5000 nm) range.

HMFGs were initially slated for optical fibre applications, because the intrinsic losses of a mid-IR fibre could in principle be lower than those of silica fibers, which are transparent only up to about 2 ìm. However, such low losses were never realized in practice, and the fragility and high cost of fluoride fibers made them less than ideal as primary candidates. Later, the utility of fluoride fibers for various other applications was discovered. These include mid-IR spectroscopy, fibre optic sensors, thermometry, and imaging. Also, fluoride fibers can be used for guided lightwave transmission in media such as YAG (yttrium aluminium garnet) lasers at 2.9 ìm, as required for medical applications (*e.g.*. ophthalmology and dentistry).

PHOSPHATE GLASS

Phosphate glass constitutes a class of optical glasses composed of metaphosphates of various metals. Instead of the SiO_4 tetrahedra observed in silicate glasses, the building block for this glass former is phosphorus pentoxide (P_2O_5), which crystallizes in at least four different forms. The most familiar polymorph (see figure) comprises molecules of P_4O_{10}. Phosphate glasses can be advantageous over silica glasses for optical fibers with a high concentration of doping rare earth ions. A mix of fluoride glass and phosphate glass is fluorophosphate glass.

CHALCOGENIDE GLASS

The chalcogens—the elements in group 16 of the periodic table—particularly sulfur (S), selenium (Se) and tellurium (Te)—react with more electropositive elements, such as silver, to form chalcogenides. These are extremely versatile compounds, in that they can be crystalline or amorphous, metallic or semiconducting, and conductors of ions or electrons. Glass containing chalcogenides can be used to make fibers for far infrared transmission.

PROCESS

Preform

Standard optical fibers are made by first constructing a large-diameter "preform" with a carefully controlled refractive index profile, and then "pulling" the preform to form the long, thin optical fibre. The preform is commonly made by three chemical vapor deposition methods: *inside vapor deposition*, *outside vapor deposition*, and *vapor axial deposition*.

With *inside vapor deposition*, the preform starts as a hollow glass tube approximately 40 centimeters (16 in) long, which is placed horizontally and rotated slowly on a lathe. Gases such as silicon tetrachloride ($SiCl_4$) or germanium tetrachloride ($GeCl_4$) are injected with oxygen in the end of the tube. The gases are then heated by means of an external hydrogen burner, bringing the temperature of the gas up to 1900 K (1600 °C, 3000 °F), where the tetrachlorides react with oxygen to produce silica or germania (germanium dioxide) particles. When the reaction conditions are chosen to allow this reaction to occur in the gas phase throughout the tube volume, in contrast to earlier techniques where the reaction occurred only on the glass surface, this technique is called *modified chemical vapor deposition (MCVD).*

The oxide particles then agglomerate to form large particle chains, which subsequently deposit on the walls of the tube as soot. The deposition is due to the large difference in temperature between the gas core and the wall causing the gas to push the particles outwards (this is known as thermophoresis). The torch is then traversed up and down the length of the tube to deposit the material evenly. After the torch has reached the end of the tube, it is then brought back to the beginning of the tube and the deposited particles are then melted to form a solid layer. This process is repeated until a sufficient amount of material has been deposited. For each layer the composition can be modified by varying the gas composition, resulting in precise control of the finished fiber's optical properties.

In outside vapor deposition or vapor axial deposition, the glass is formed by *flame hydrolysis*, a reaction in which silicon tetrachloride and germanium tetrachloride are oxidized by reaction with water (H_2O) in an oxyhydrogen flame. In outside vapor deposition the glass is deposited onto a solid rod, which is removed before further processing. In vapor axial deposition, a short *seed rod* is used, and a porous preform, whose length is not limited by the size of the source rod, is built up on its end. The porous preform is consolidated into a transparent, solid preform by heating to about 1800 K (1500 °C, 2800 °F).

Typical communications fibre uses a circular preform. For some applications such as double-clad fibers another form is preferred. In fibre lasers based on double-clad fibre, an asymmetric shape improves the filling factor for laser pumping. Because of the surface tension, the shape is smoothed during the drawing process, and the shape of the resulting fibre does not reproduce the sharp edges of the preform. Nevertheless, careful polishing of the preform is important, since any defects of the preform surface affect the optical and mechanical properties of the resulting fibre. In particular, the preform for the test-fibre shown in the figure was not polished well, and cracks are seen with the confocal optical microscope.

Drawing

The preform, however constructed, is placed in a device known as a drawing tower, where the preform tip is heated and the optical fibre is pulled out as a string. By measuring the resultant fibre width, the tension on the fibre can be controlled to maintain the fibre thickness.

COATINGS

The light is guided down the core of the fibre by an optical cladding with a lower refractive index that traps light in the core through total internal reflection.

The cladding is coated by a buffer that protects it from moisture and physical damage. The buffer coating is what gets stripped off the fibre for termination or splicing. These coatings are UV-cured urethane acrylate composite materials applied to the outside of the fibre during the drawing process. The coatings protect the very delicate strands of glass fibre—about the size of a human hair—and allow it to survive the rigours of manufacturing, proof testing, cabling and installation.

Today's glass optical fibre draw processes employ a dual-layer coating approach. An inner primary coating is designed to act as a shock absorber to minimize attenuation caused by microbending. An outer secondary coating protects the primary coating against mechanical damage and acts as a barrier to lateral forces. Sometimes a metallic armor layer is added to provide extra protection. These fibre optic coating layers are applied during the fibre draw, at speeds approaching 100 kilometers per hour (60 mph). Fibre optic coatings are applied using one of two methods: *wet-on-dry* and *wet-on-wet*. In wet-on-dry, the fibre passes through a primary coating application, which is then UV cured—then through the secondary coating application, which is subsequently cured. In wet-on-wet, the fibre passes through both the primary and secondary coating applications, then goes to UV curing.

Fibre optic coatings are applied in concentric layers to prevent damage to the fibre during the drawing application and to maximize fibre strength and microbend resistance. Unevenly coated fibre will experience non-uniform forces when the coating expands or contracts, and is susceptible to greater signal attenuation. Under proper drawing and coating processes, the coatings are concentric around the fibre, continuous over the length of the application and have constant thickness. Fibre optic coatings protect the glass fibers from scratches that could lead to strength degradation. The combination of moisture and scratches accelerates the aging and deterioration of fibre strength. When fibre is subjected to low stresses over a long period, fibre fatigue can occur. Over time or in extreme conditions, these factors combine to cause microscopic flaws in the glass fibre to propagate, which can ultimately result in fibre failure.

Three key characteristics of fibre optic waveguides can be affected by environmental conditions: strength, attenuation and resistance to losses caused by microbending. External fibre optic coatings protect glass optical fibre from environmental conditions that can affect the fiber's performance and long-term durability. On the inside, coatings ensure the reliability of the signal being carried and help minimize attenuation due to microbending.

PRACTICAL ISSUES

CABLE CONSTRUCTION

In practical fibers, the cladding is usually coated with a tough resin *buffer* layer, which may be further surrounded by a *jacket* layer, usually glass. These layers add strength to the fibre but do not contribute to its optical wave guide properties. Rigid fibre assemblies sometimes put light-absorbing ("dark") glass between the fibers, to prevent light that leaks out of one fibre from entering another. This reduces cross-talk between the fibers, or reduces flare in fibre bundle imaging applications.

Modern cables come in a wide variety of sheathings and armor, designed for applications such as direct burial in trenches, high voltage isolation, dual use as power lines, installation in conduit, lashing to aerial telephone poles, submarine installation, and insertion in paved streets. The cost of small fibre-count pole-mounted cables has greatly decreased due to the high demand for fibre to the home (FTTH) installations in Japan and South Korea.

Fibre cable can be very flexible, but traditional fiber's loss increases greatly if the fibre is bent with a radius smaller than around 30 mm. This creates a problem when the cable is bent around corners or wound around a spool, making FTTX installations more complicated. "Bendable fibers", targeted towards easier installation in home environments, have been standardized as ITU-T G.657. This type of fibre can be bent with a radius as low as 7.5 mm without adverse impact. Even more bendable fibers have been developed. Bendable fibre may also be resistant to fibre hacking, in which the signal in a fibre is surreptitiously monitored by bending the fibre and detecting the leakage.

Another important feature of cable is cable's ability to withstand horizontally applied force. It is technically called max tensile strength defining how much force can be applied to the cable during the installation period.

Some fibre optic cable versions are reinforced with aramid yarns or glass yarns as intermediary strength member. In commercial terms, usage of the glass yarns are more cost effective while no loss in mechanical durability of the cable. Glass yarns also protect the cable core against rodents and termites.

TERMINATION AND SPLICING

Optical fibers are connected to terminal equipment by optical fibre connectors. These connectors are usually of a standard type such as *FC*, *SC*, *ST*, *LC*, *MTRJ*, or *SMA*, which is designated for higher power transmission.

Optical fibers may be connected to each other by connectors or by *splicing*, that is, joining two fibers together to form a continuous optical waveguide. The generally accepted splicing method is arc fusion splicing, which melts the fibre ends together with an electric arc. For quicker fastening jobs, a "mechanical splice" is used.

Fusion splicing is done with a specialized instrument that typically operates as follows: The two cable ends are fastened inside a splice enclosure that will protect the splices, and the fibre ends are stripped of their protective polymer coating (as well as the more sturdy outer jacket, if present). The ends are *cleaved* (cut) with a precision cleaver to make them perpendicular, and are placed into special holders in the splicer. The splice is usually inspected via a magnified viewing screen to check the cleaves before and after the splice. The splicer uses small motors to align the end faces together, and emits a small spark between electrodes at the gap to burn off dust and moisture. Then the splicer generates a larger spark that raises the temperature above the melting point of the glass, fusing the ends together permanently. The location and energy of the spark is carefully controlled so that the molten core and cladding do not mix, and this minimizes optical loss. A splice loss estimate is measured by the splicer, by directing light through the cladding on one side and measuring the light leaking from the cladding on the other side. A splice loss under 0.1 dB is typical. The complexity of this process makes fibre splicing much more difficult than splicing copper wire.

Mechanical fibre splices are designed to be quicker and easier to install, but there is still the need for stripping, careful cleaning and precision cleaving. The fibre ends are aligned and held together by a precision-made sleeve, often using a clear index-matching gel that enhances the transmission of light across the joint. Such joints typically have higher optical loss and are less robust than fusion splices, especially if the gel is used. All splicing techniques involve installing an enclosure that protects the splice.

Fibers are terminated in connectors that hold the fibre end precisely and securely. A fibre-optic connector is basically a rigid cylindrical barrel surrounded by a sleeve that holds the barrel in its mating socket. The mating mechanism can be *push and click*, *turn and latch* (*bayonet mount*), or *screw-in* (*threaded*). A typical connector is installed by preparing the fibre end and inserting it into the rear of the connector body. Quick-set adhesive is usually used to hold the fibre securely, and a strain relief is secured to the rear. Once the adhesive sets, the fiber's end is polished to a mirror finish. Various polish profiles are used, depending on the type of fibre and the application. For single-mode fibre, fibre ends are typically polished with a slight curvature that makes the mated connectors touch only at their cores. This is called a *physical contact* (PC) polish. The curved surface may be polished at an angle, to make an *angled physical contact (APC)* connection. Such connections have higher loss than PC connections, but greatly reduced back reflection, because light that reflects from the angled surface leaks out of the fibre core. The resulting signal strength loss is called *gap loss*. APC fibre ends have low back reflection even when disconnected.

In the 1990s, terminating fibre optic cables was labour-intensive. The number of parts per connector, polishing of the fibers, and the need to oven-bake the epoxy in each connector made terminating fibre optic cables difficult. Today, many connectors types are on the market that offer easier, less labour-intensive ways of terminating cables. Some of the most popular connectors are pre-polished at the factory, and include a gel inside the connector. Those two steps help save money on labour, especially on large projects. A cleave is made at a required length, to get as close to the polished piece already inside the connector. The gel surrounds the point where the two pieces meet inside the connector for very little light loss.

FREE-SPACE COUPLING

It is often necessary to align an optical fibre with another optical fibre, or with an optoelectronic device such as a light-emitting diode, a laser diode, or a modulator. This can involve either carefully aligning the fibre and placing it in contact with the device, or can use a lens to allow coupling over an air gap. In some cases the end of the fibre is polished into a curved form that makes it act as a lens. Some companies can even shape the fibre into lenses by cutting them with lasers.

In a laboratory environment, a bare fibre end is coupled using a fibre launch system, which uses a microscope objective lens to focus the light down to a fine point. A precision translation stage (micro-positioning table) is used to move the lens, fibre, or device to allow the coupling efficiency to be optimized. Fibers with a connector on the end make this process much simpler: the connector is simply plugged into a pre-aligned fiberoptic collimator, which contains a lens that is either accurately positioned with respect to the fibre, or is adjustable. To achieve the best injection efficiency into single-mode fibre, the direction, position, size and divergence of the beam must all be optimized. With good beams, 70 to 90 per cent coupling efficiency can be achieved.

With properly polished single-mode fibers, the emitted beam has an almost perfect Gaussian shape—even in the far field—if a good lens is used. The lens needs to be large enough to support the full numerical aperture of the fibre, and must not introduce aberrations in the beam. Aspheric lenses are typically used.

FIBRE FUSE

At high optical intensities, above 2 megawatts per square centimeter, when a fibre is subjected to a shock or is otherwise suddenly damaged, a *fibre fuse* can occur. The reflection from the damage vaporizes the fibre immediately before the break, and this new defect remains reflective so that the damage propagates back towards the transmitter at 1–3 meters per second (4–11 km/h, 2–8 mph). The open fibre control system, which ensures laser eye safety in the event of a broken fibre, can also effectively halt propagation of the fibre fuse. In situations, such as undersea cables, where high power levels might be used without the need for open fibre control, a "fibre fuse" protection device at the transmitter can break the circuit to keep damage to a minimum.

ALL-DIELECTRIC SELF-SUPPORTING CABLE

All-dielectric self-supporting (ADSS) cable is a type of optical fibre cable that is strong enough to support itself between structures without using conductive metal elements. It is used by electrical utility companies as a communications medium, installed along existing overhead transmission lines and often sharing the same support structures as the electrical conductors.

ADSS is an alternative to OPGW and OPAC with lower installation cost. The cables are designed to be strong enough to allow lengths of up to 700 metres to be installed between support towers. ADSS cable is designed to be light weight and small in diameter to reduce the load on tower structures due to cable weight, wind, and ice.

In the design of the cable, the internal glass optical fibers are supported with no strain, to maintain low optical loss throughout the life of the cable. The cable is jacketed to prevent moisture from degrading the fibers. The jacket also protects the polymer strength elements from the effect of solar ultraviolet light.

Using single-mode fibers and light wavelengths of either 1310 or 1530 nanometres, circuits up to 100 km long are possible without repeaters. A single cable can carry as many as 144 fibers.

CONSTRUCTION DETAILS

No metal wires are used in an ADSS cable. Optical fibers are either supported in loose buffer tubes, or arranged in a ribbon configuration. To prevent strain on the fibers, most types provide the fibres with excess slack length compared to the length of the supporting member.

Several designs of cable are used. In one type, small plastic tubes each containing up to eight fibers are filled with a gel compound, and stranded around plastic rods. The hydrocarbon gel compound prevents water from wicking along the length of the cable. The inner structure is wrapped with polyester tapes and encased in a pultruded glass reinforced plastic tube, which provides the structural strength for the cable. An outer sheath is applied to protect the strength member from ultraviolet light and from damage due to leakage current.

Another type uses a glass reinforced plastic rod with a notch in one side, which carries one or two ribbons of glass fibers. The notch is filled with water-resistant gel and covered with an extruded plastic filler strip. The whole assembly is enclosed in a sunlight and tracking resistant sheath. This design is not as flexible as the symmetrical cylindrical design above, but provides good access to the fibers at joints and is resistant to crushing. At full working load the optical fibers may experience up to 0.25 per cent strain, which is undesirable for long fibre life.

A third type gets its strength from aramid fibre yarns, which are coated to prevent water wicking. The aramid yarn strength member surrounds a core made up of multiple buffer tubes, each with a fibre, all surrounding a plastic core. The outer sheath provides protection from water and sunlight. This design does not have as great resistance to crushing.

Another type of design uses four glass-reinforced plastic strength member strands, and loose buffer tubes cabled into an assembly and protected by a jacket. This is more flexible than the outer cylinder type first described,but does not provide as much protection to the fibers.

ADSS cables made by BICC with 24 or 48 fibers in the above designs range from 220 to 284 kg/kilometre in weight and are between 13 and 16 mm outside jacket diameter. These cables can support between 20 and 25 kilonewtons of tension.

ACCESSORIES AND INSTALLATION

Fittings used with ADSS cable may be tension type, used at dead-ends where the cable terminates or changes direction, or may be suspension type, only holding the weight of a span with tension transmitted through the next span of cable. Reinforcing rods are used on either side of a support and at dead-ends. Wind-induced aeolian vibration is a factor since ADSS cables have light weight, relatively high tension, and little self-damping. Anti-vibration dampers are installed on each span near the support points. Accessories must not be clamped directly to the cable but instead over reinforcing rods, to protect the cable from electrical and mechanical damage. Termination boxes are used to enclose and protect splices between the ADSS cable and "inside plant" cable runs.

ADSS cable can be installed using live-line methods on an energized transmission line. Fibre cables are generally supported on the lower cross-arms of the tower, which provides good clearance to the ground. When the fibers are installed in the middle of a tower, the fibre cable is unlikely to hit energized conductors. Lower weights and forces are used for installation, compared with metallic cables, so lighter equipment can be used. Installation technique is similar to installing overhead conductors, with care taken to prevent excessively tight bending of the cable, and adjustment of the sag of individual spans as for metallic cables.

APPLICATION ISSUES

Cables must be designed for the worst-case combinations of temperature, ice load, and wind. An installed cable must not sag so low that it can be damaged by traffic under the line. Cables must be installed so that "galloping" caused by sustained high wind does not cause the cable to hit conductors, and so that "sleet jump" caused by ice melting off the cable does not cause impacts. The water-blocking gel must not become so stiff at low temperature that the fibers are put under strain.

Transmission lines are exposed to damage by gunfire, especially in rural areas. Shotgun pellets can sever fibers or damage the sheath allowing water into the cable. Adding a ballistic shield element to the cable makes it larger and heavier and may not be economically feasible. The utility may factor gunshot damage into reliability calculations for the system.

Glass under tension and exposed to acid environments loses strength; this applies to both the optical fibers and the glass reinforcement of polymers. The cable jacket and gel coating of fibres provides protection from chemical attack.

The ADSS cable is suspended in the electrical field due to the phase conductors; this varies from a maximum at mid-span to zero at the grounded metal supports of the cable. In dry conditions, no current flows on the jacket of the cable, but moisture reduces the jacket insulation. Uneven distribution of moisture can result in formation of high-resistance "dry bands" which have a relatively high voltage across them. Dry bands tend to form at the supports. Voltage across the dry band can cause carbon tracks to form and erosion of the jacket material. If the voltage across the dry band is high enough, an arc may form which can damage the jacket. Dry-band arcing is more likely for cables installed under higher transmission voltage lines (220 kV and above). Even a few incidents of arcing along a dry band can cause severe permanent damage to the jacket, leading to subsequent failure of the cable. Relatively low sustained arc currents of a few milliamperes can cause eventual aging degradation of the cable. The magnitude of current available in an arc (and probability of damage) depends on the geometry of the installation and is not simply correlated with the voltage of the transmission line. Wetting conditions near industrial plants or saltwater will have more severe effect on the jacket resistance than in freshwater rain or fog. While one measure to protect cables from dry-banding damage incorporates a semiconducting layer to equalize potential, a more feasible method uses externally applied protection within 50 metres of each support, since this is the area most susceptible to damage.

ALL-SILICA FIBRE

All-silica fibre, or silica-silica fibre, is an optical fibre whose core and cladding are made of silica glass. The refractive index of the core glass is higher than that of the cladding. These fibers are typically step-index fibers. The cladding of an all-silica fibre should not be confused with the polymer overcoat of the fibre.

CLEAR CURVE

Clear Curve is Corning's brand name for a new optical fibre that can be bent around short-radius curves without losing its signal. It is constructed with a conventional fibre on the inside, surrounded by a cladding containing a new nanostructured reflector. ClearCurve is hundreds of times more flexible than conventional optical cable, transmitting high-quality signals even when wrapped around small objects like a pen, where a conventional cable would lose the signal completely.

Although originally introduced to serve the needs of pulling fibre in apartment buildings and other high-density units where conventional fibre is too inflexible, in 2009 Intel announced their intention to use it as the basis of a new computer interconnect system code-named Light Peak. ClearCurve's small size and high bandwidth capabilities offer great improvements over existing copper wiring in this role, and Intel is positioning Light Peak as a truly universal bus that can carry any existing traffic over a single cable.

CONVENTIONAL FIBRE

Conventional optical fibre consists of a thin inner cylindrical core of glass or plastic with a similar material layered in a thin coating around it. Slight differences in the index of refraction between the two layers causes total internal reflection, trapping a light beam inside the inner core. This process is limited to a critical angle; when the light beam approaches the interface at a shallow angle most of it will be reflected, but as it gets closer to the critical angle more and more will travel through the interface and be lost.

The critical angle depends on the relative difference in index of refraction, larger differences will increase the critical angle and trap more light. However, changing the index of refraction in most materials generally changes its mechanical properties too, which means that different types of cables are used for different purposes. Cables intended to be highly efficient over long runs are generally less flexible, while those that require higher flexibility are generally only useful for shorter distances. Even cables designed to be flexible, like TOSLINK, are less flexible than a similar sized braided copper cable.

In order to keep the fibers as straight as possible, most high-performance optical cables use a form of armour that resists tight bending. This normally takes the form of a helical winding, similar to BX cable, or a series of straight fibers running parallel to the core. Since the armour is fairly large, the cables normally carry a number of fibers inside. The resulting armoured bundle is then surrounded in some sort of environmental cladding, typically plastic. The bundle is about the size of a conventional power cable that you might find on an electrical appliance, but much less flexible.

FIBRE TO THE HOME

Optical cabling has long formed the backbone of major terrestrial networks, delivering signals over long distances. The signals are then converted into other forms at the company end offices, and distributed from there in some other form, typically telephone wiring or coax cable in the case of cable television. The multi-fibre armoured cable is well suited to this role.

Since the 1990s there has been an ongoing effort to supply fibre to the home (FTTH). Using fibre to deliver signals all the way to the home provides the same advantages as it does on the longer hauls, namely much higher bandwidth, lower costs, and less interference with other sources. However, given the deliberate lack of flexibility of the cable, these installations generally end in a utility room where they are converted to copper for distribution within the home.

While this sort of installation is useful for individual dwellings, it is less useful in large multi-unit dwellings. Corning estimates that an apartment installation would require an average of twelve right angle turns between the distribution point and the units. Conventional fibre would lose the signal after one or two such bends, making it useless in this role. As is the case for individual homes, the fibre can be converted to copper for the last section of delivery, but the longer runs demand much higher performance, larger cable. Finding room to run these cables in an existing structure may not be possible.

CLEARCURVE

ClearCurve fibers are constructed in a fashion similar to existing cables, starting with a traditional glass fibre in the center. ClearCurve then adds a third layer to the sandwich, a plastic sheath that is infused with microscopic reflectors. Light that passes through the conventional interface has a second chance to be reflected back into the center of the fibre. In the corners of tight bends, the reflectors serve to increase the amount of signal retained within the cable, allowing ClearCurve to be hundreds of times more flexible than conventional cables. A thin environmental sheath is added on the outside.

Unlike conventional fibers, ClearCurve does not have to be held straight, and therefore eliminates the armour. Lacking armour, there is no lower limit to the size of a ClearCurve cable, which can be as small as a single fibre, although normally they contain two fibres, one upstream and one down. Two-fibre ClearCurve cables are smaller than the wire on a typical computer mouse, yet the high-performance single-mode versions carry 25 Gbit/s over long lengths.

In a video demonstration, Corning showed a ClearCurve drop cable being wrapped dozens of times around a small metal rod, and suffering almost no signal loss and providing a perfect video feed. A conventional cable wrapped around the same rod completely lost the signal after only two turns.

FTTH uses

ClearCurve is the end result of a Corning research project looking for products better tailored to the fibre to the home market. Running since 1988 at their Sullivan Park research center in New York, Corning announced ClearCurve at a press event on 19 September 2007 and showed it publicly at the FTTH Conference later that month.

Using ClearCurve, a FTTH installation can use existing armoured cabling to deliver the signal to a utility room, then connect individual ClearCurve cables to the fibers in the bundle for distribution within the building. This sort of installation dramatically simplifies the overall complexity of FTTH wiring in multi-unit dwellings, eliminating both the large coax cabling and the need to convert formats from light to electrical. Users were quickly forthcoming; announced in September, only a month later an official press release announced that Connexion Technologies would be using ClearCurve on 30 November 2007. Since then many additional partners have been announced.

Computer bus uses

The single-mode fibers used in conversional telecommunications applications have high performance, but require expensive light sources and highly accurate mechanical positioning in order to gather light from them. In comparison, multi-mode have wider cores that are easier to connect to and can be effectively driven by lower-cost devices like solid-state IR lasers or vertical-cavity surface-emitting lasers (VCSELs).

Multi-mode fibre found some uses in high-performance computing applications, notably the Fibre Channel system for high speed disks and some parallel computing interconnection systems. However, the relatively inflexible cables made them less useful in general roles, where braided copper wiring remains widespread. Fibre has found one consumer use, the TOSLINK cable used in digital audio applications. This role uses lower quality multi-mode plastic fibre with limited bandwidth, about 125 Mbit/s, driven by red LEDs. However, advances in computers have demanded ever increasing bandwidth, and modern computer bus systems are quickly reaching their limits. There was some discussion of moving to optical fibre for the USB3 standard, but the decision was made to move ahead with copper.

Corning announced a multi-mode version of ClearCurve cabling on 13 January 2009. It has greater bandwidth than any common copper wiring, and is at least as flexible as copper wiring able to carry the same amount of data. Although it was mentioned only in passing, Intel's new Light Peak interconnection system uses ClearCurve cabling as its basis. Light Peak uses a two-fibre cable running at 10 Gbit/s in both directions. Unlike most optical connection systems, Light Peak is being designed to allow daisy-chaining and supply power through a set of coaxial copper wires.

DISPERSION-SHIFTED FIBRE

Dispersion-shifted fibre (DSF) is a type of optical fibre made to optimize both low dispersion and low attenuation. Dispersion Shifted Fibre is a type of single-mode optical fibre with a core-clad index profile tailored to shift the zero-dispersion wavelength from the natural 1300 nm in silica-glass fibers to the minimum-loss window at 1550 nm. The group velocity or *intramodal* dispersion which dominates in single-mode fibers includes both material and waveguide dispersion. Waveguide dispersion can be made more negative by changing the index profile and thus be used to offset the fixed material dispersion, shifting or flattening the overall intramodal dispersion. This is advantageous because it allows a communication system to possess both low dispersion and low attenuation. However, when used in wavelength division multiplexing systems, dispersion-shifted fibers can suffer from four-wave mixing which causes intermodulation of the independent signals. As a result nonzero dispersion shifted fibre is often used. Dispersion-shifted fibre is specified in ITU-T G.653.

DISTRIBUTED ACOUSTIC SENSING

Rayleigh scattering based distributed acoustic sensing (DAS) systems use fibre optic cables to provide distributed strain sensing. In DAS, the optical fibre cable becomes the sensing element and measurements are made, and in part processed, using an attached optoelectronic device. Such a system allows acoustic frequency strain signals to be detected over large distances and in harsh environments.

FUNDAMENTALS OF RAYLEIGH SCATTER BASED FIBRE OPTIC SENSING

In Rayleigh scatter based distributed fibre optic sensing, a coherent laser pulse is sent along an optic fibre, and scattering sites within the fibre cause the fibre to act as a distributed interferometer with a gauge length approximately equal to the pulse length. The intensity of the reflected light is measured as a function of time after transmission of the laser pulse. When the pulse has had time to travel the full length of the fibre and back, the next laser pulse can be sent along the fibre. Changes in the reflected intensity of successive pulses from the same region of fibre are caused by changes in the optical path length of that section of fibre. This type of system is very sensitive to both strain and temperature variations of the fibre and measurements can be made simultaneously at all sections of the fibre.

CAPABILITIES OF RAYLEIGH-BASED SYSTEMS

Maximum range

The optical pulse is attenuated as it propagates along the fibre. For a single mode fibre operating at 1550 nm, a typical attenuation is 0.2 dB/km. Since the light must make a double pass along each section of fibre, this means each 1 km causes a total loss of 0.4dB.

The maximum range of the system occurs when the amplitude of the reflected pulse becomes so low it is impossible to obtain a clear signal from it. It is not possible to counteract this effect by increasing the input power because above a certain level this will induce nonlinear optical effects which will disrupt the operation of the system. Typically the maximum range that can be measured is around 40–50 km.

Strain resolution

The minimum value of strain that can be measured depends on the carrier to noise ratio of the returning optical signal. The carrier level is largely determined by the amplitude of the optical signal while the noise is a combination of that from a variety of sources including laser noise, electronic noise and detector noise.

Spatial resolution and spatial sampling period

The spatial resolution is mainly determined by the duration of the transmitted pulse, with a 100ns pulse giving 10m resolution being a typical value. The amount of reflected light is proportional to the pulse length so there is a trade-off between spatial resolution and maximum range. To improve the maximum range it would be desirable to use a longer pulse length to increase the reflected light level but this leads to a larger spatial resolution. In order for two signals to be independent they must be obtained from two points on the fibre that are separated by at least the spatial resolution. It is possible to obtain samples at separations less than the spatial resolution and although this produces signals that are not independent of each other such an approach does offer advantages in some applications. The separation between the sampling points is sometimes referred to as the spatial sampling period.

Acquisition rate

Before the next laser pulse can be transmitted the previous one must have had time to travel to the far end of the fibre and for the reflections from there to return, otherwise reflections would be returning from different sections of the fibre at the same time and the system would not operate properly. For a fibre 50 km long the maximum pulse rate is just over 2 kHz. Therefore strains can be measured which vary at frequencies up to the Nyquist frequency of 1 kHz. Shorter fibers clearly enable higher acquisition rates.

Temperature measurements

Although the system is sensitive to both temperature and strain variations these can often be separated as those due to temperature tend to occur at a lower frequency range than strain. Unlike other distributed fibre techniques such as those based on Brillouin or Raman scatter, distributed acoustic sensing is only able to detect changes in temperature rather than its absolute value.

COMPARISON WITH OTHER FIBRE OPTIC DISTRIBUTED SENSING TECHNIQUES

Distributed Acoustic Sensing relies on light which is Rayleigh backscattered from small variations in the refractive index of the fibre. The backscattered light has the same frequency as the transmitted light. There are a number of other distributed fibre sensing techniques that rely on different scattering mechanisms and can be used to measure other parameters. Brillouin scatter occurs due to the interaction between the light and acoustic phonons travelling in the fibre. As the light is scattered by a moving phonon its frequency is shifted by the Doppler effect by around 10 GHz. Light is generated at both at above (anti-Stokes shift) and below (Stokes shift) the original optical frequency. The intensity and frequency shifts of the two components are dependent on both temperature and strain and by measuring the shifts, absolute values of the two parameters can be calculated using a Distributed Temperature and Strain Sensing (DTSS) system. Brillouin scatter is much weaker than Rayleigh scatter and so the reflections from a number of pulses must be summed together to enable the measurements to be made. Therefore the maximum frequency at which changes can be measured using Brillouin scatter is typically a few 10's of Hz. Raman scatter occurs when light is scattered in interaction with molecular vibrations in the fibre. As with Brillouin scattering both Stokes and anti-Stokes components are produced and these are shifted from the wavelength of the incident light by several tens of nanometers. By measuring the ratio in intensity between the Stokes and anti-Stokes components an absolute value of temperature can be measured by a distributed temperature sensing (DTS) system. The larger wavelength shifts compared to Brillouin scatter mean that it is easier to separate the scattered Raman light from the un-shifted Rayleigh scattered component. However the intensity of the Raman scatter is even lower than the Brillouin scatter and so it is normally necessary to average for many seconds or even minutes in order to get reasonable results. Therefore Raman based systems are only suitable for measuring slowly varying temperatures.

APPLICATIONS

The sensitivity and speed of Rayleigh based sensing allows distributed acoustic monitoring over distances of up to 100 km from each laser source. With suitable analysis software, continuous monitoring of pipelines for unwanted interference, as well as leaks or flow irregularities is possible. Roads, borders, perimeters etc. can be monitored for unusual activity with the position of the activity being determined to within approximately 10 metres. Due to the ability of the optic fibre to operate in harsh environments, the technology can also be used in oil well monitoring applications, allowing real-time information on the state of the well to be determined.

DOUBLE-CLAD FIBRE

Double-clad fibre (DCF) is a class of optical fibre with a structure consisting of three layers of optical material instead of the usual two. The inner-most layer is called the *core*. It is surrounded by the *inner cladding*, which is surrounded by the *outer cladding*. The three layers are made of materials with different refractive indices.

There are two different kinds of double-clad fibers. The first was developed early in optical fibre history with the purpose of engineering the dispersion of optical fibers. In these fibers, the core carries the majority of the light, and the inner and outer cladding alter the waveguide dispersion of the core-guided signal. The second kind of fibre was developed in the late 1980s for use with high power fibre amplifiers and fibre lasers. In these fibers, the core is doped with active dopant material; it both guides and amplifies the signal light. The inner cladding and core together guide the pump light, which provides the energy needed to allow amplification in the core. In these fibers, the core has the highest refractive index and the outer cladding has the lowest. In most cases the outer cladding is made of a polymer material rather than glass.

DISPERSION-COMPENSATING FIBRE

In double-clad fibre for dispersion compensation, the inner cladding layer has lower refractive index than the outer layer. This type of fibre is also called *depressed-inner-cladding fibre* and *W-profile fibre* (from the fact that a symmetrical plot of its refractive index profile superficially resembles the letter W).

This type of double-clad fibre has the advantage of very low microbending losses. It also has two zero-dispersion points, and low dispersion over a much wider wavelength range than standard singly clad fibre. Since the dispersion of such double-clad fibers can be engineered to a great extent, these fibers can be used for the compensation of chromatic dispersion in optical communications and other applications.

FIBRE FOR AMPLIFIERS AND FIBRE LASERS

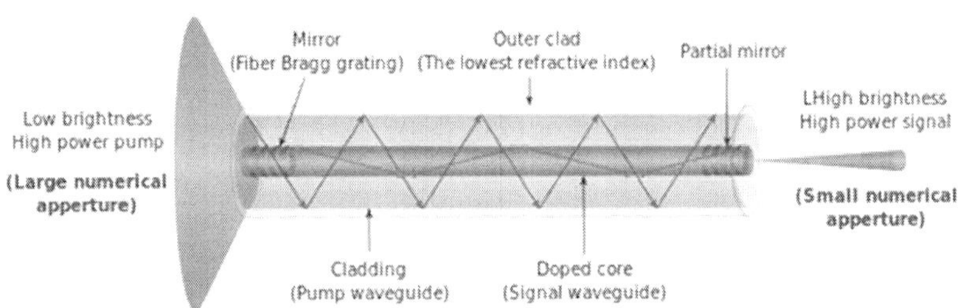

Schematic diagram of cladding-pumped double-clad fiber laser

In modern double-clad fibers for high power fibre amplifiers and lasers, the inner cladding has a higher refractive index than the outer cladding. This enables the inner cladding to guide light by total internal reflection in the same way the core does, but for a different range of wavelengths. This allows diode lasers, which have high power but low brightness, to be used as the optical pump source. The pump light can be easily coupled into the large inner cladding, and propagates through the inner cladding while the signal propagates in the smaller core. The doped core gradually absorbs the cladding light as it propagates, driving the amplification process. This pumping scheme is often called *cladding pumping*, which is an alternative to the conventional *core pumping*, in which the pump light is coupled into the small core. The invention of cladding pumping by a Polaroid fibre research team (H. Po, *et al.*) revolutionized the design of fibre amplifiers and lasers. Using this method, modern fibre lasers can produce continuous power up to several kilowatts, while the signal light in the core maintains near diffraction-limited beam quality.

The shape of the cladding is very important, especially when the core diameter is small compared to the size of the inner cladding. Circular symmetry in a double-clad fibre seems to be the worst solution for a fibre laser; in this case, many modes of the light in the cladding miss the core and hence cannot be used to pump it. In the language of geometrical optics, most of the rays of the pump light do not pass through the core, and hence cannot pump it. Ray tracing, simulations of the paraxial propagation and mode analysis give similar results.

Chaotic fibers

In general, modes of a waveguide have "scars", which correspond to the classical trajectories. The scars may avoid the core, then the mode is not coupled, and it is vain to excite such a mode in the double-clad fibre amplifier. The scars can be distributed more or less uniformly in so-called chaotic fibers have more complicated cross-sectional shape and provide more uniform distribution of intensity in the inner cladding, allowing efficient use of the pump light. However, the scarring takes place even in chaotic fibers.

Spiral shape

An almost-circular shape with small spiral deformation seems to be the most efficient for chaotic fibers. In such a fibre, the angular momentum of a ray increases at each reflection from the smooth wall, until the ray hits the "chunk", at which the spiral curve is broken (see figure at right). The core, placed in vicinity of this chunk, is intercepted more regularly by all the rays compared to other chaotic fibers. This behaviour of rays has an analogy in wave optics. In the language of modes, all the modes have non-zero derivative in vicinity of the chunk, and cannot avoid the core if it is placed there. One example of modes is shown in the figure below and to the right. Although some of modes show scarring and wide voids, none of these voids cover the core.

The property of DCFs with spiral-shaped cladding can be interpreted as conservation of angular momentum. The square of the derivative of a mode at the boundary can be interpreted as pressure. Modes (as well as rays) touching the spiral-shaped boundary transfer some angular momentum to it. This transfer of angular momentum should be compensated by pressure at the chunk. Therefore, no one mode can avoid the chunk. Modes can show strong scarring along the classical trajectories (rays) and wide voids, but at least one of scars should approach the chunk to compensate for the angular momentum transferred by the spiral part.

The interpretation in terms of angular momentum indicates the optimum size of the chunk. There is no reason to make the chunk larger than the core; a large chunk would not localize the scars sufficiently to provide coupling with the core. There is no reason to locaize the scars within an angle smaller than the core: the small derivative to the radius makes the manufacturing less robust; the larger $R'(\phi)$ is, the larger the fluctuations of shape that are allowed without breaking the condition $R'(\phi) > 0$. Therefore, the size of the chunk should be of the same order as the size of the core.

More rigorously, the property of the spiral-shaped domain follows from the theorem about boundary behaviour of modes of the Dirichlet Laplacian. Although this theorem is formulated for the core-less domain, it prohibits the modes avoiding the core. A mode avoiding the core, then, should be similar to that of the core-less domain. Stochastic optimization of the cladding shape confirms that an almost-circular spiral realizes the best coupling of pump into the core.

Filling factor

The efficiency of absorption of pumping energy in the fibre is an important parameter of a double-clad fibre laser. In many cases this efficiency can be approximated with

$$1 - \exp\left(-F \frac{\pi r^2}{S} \alpha L \right),$$

where

S is the cross-sectional area of the cladding

r is the radius of the core (which is taken to be circular)

α is the absorption coefficient of pump light in the core

L is the length of the double-clad fibre, and

F is a dimensionless adjusting parameter, which is sometimes called the "filling factor"; $0 < F < 1$.

The filling factor may depend on the initial distribution of the pump light, the shape of the cladding, and the position of the core within it.

The exponential behaviour of the efficiency of absorption of pump in the core is not obvious. One could expect that some modes of the cladding (or some rays) are better coupled to the core than others; therefore, the "true" dependence could be a combination of several exponentials. Only comparison with simulations justifies this approximation, as shown in the figure above and to the right. In particular, this approximation does not work for circular fibers. For chaotic fibers, F approaches unity. The value of f F can be estimated by numerical analysis with propagation of waves, expansion by modes or by geometrical optics ray tracing, and values 0.8 and 0.9 are only empirical adjusting parameters, which provide good agreement of the simple estimate with numerical simulations for two specific classes of double-clad fibers: circular offset and rectangular. Obviously, the simple estimate above fails when the offset parameter becomes small compared to the size of cladding.

The filling factor Fapproaches unity especially quickly in the spiral-shaped cladding, due to the special boundary behaviour of the modes of the Dirichlet Laplacian. Designers of double-clad fibre look for a reasonable compromise between the optimized shape (for the efficient couplung of pump into the core) and the simplicity of the manufacturing of the preform used to draw the fibers.

The power scaling of a fibre laser is limited by unwanted nonlinear effects such as stimulated Brillouin scattering and stimulated Raman scattering. These effects are minimized when the fibre laser is short. For efficient operation, however, the pump should be absorbed in the core along the short length; the estimate above applies in this optimistic case. In particular, the higher the step in refractive index from inner to outer cladding, the better-confined the pump is. As a limiting case, the index step can be of order of two, from glass to air. The estimate with filling factor gives an estimate of how short an efficient double-clad fibre laser can be, due to reduction in size of the inner cladding.

Alternative structures

For good cladding shapes the filling factor F, defined above, approaches unity; the following enhancement is possible at various kinds of tapering of the cladding; non-conventional shapes of such cladding are suggested. Planar waveguides with an active gain medium take an intermediate position between conventional solid-state lasers and double-clad fibre lasers. The planar waveguide may confine a multi-mode pump and a high-quality signal beam, allowing efficient coupling of the pump, and diffraction-limited output.

FIBRE BRAGG GRATING

A fibre Bragg grating (FBG) is a type of distributed Bragg reflector constructed in a short segment of optical fibre that reflects particular wavelengths of light and transmits all others. This is achieved by creating a periodic variation in the refractive index of the fibre core, which generates a wavelength specific dielectric mirror. A fibre Bragg grating can therefore be used as an inline optical filter to block certain wavelengths, or as a wavelength-specific reflector.

The first in-fibre Bragg grating was demonstrated by Ken Hill in 1978. Initially, the gratings were fabricated using a visible laser propagating along the fibre core. In 1989, Gerald Meltz and colleagues demonstrated the much more flexible transverse holographic inscription technique where the laser illumination came from the side of the fibre. This technique uses the interference pattern of ultraviolet laser light to create the periodic structure of the fibre Bragg grating.

MANUFACTURE

Fibre Bragg gratings are created by "inscribing" or "writing" systematic (periodic or aperiodic) variation of refractive index into the core of a special type of optical fibre using an intense ultraviolet (UV) source such as a UV laser. Two main processes are used: *interference* and *masking*. The method that is preferable depends on the type of grating to be manufactured. Normally a germanium-doped silica fibre is used in the manufacture of fibre Bragg gratings. The germanium-doped fibre is photosensitive, which means that the refractive index of the core changes with exposure to UV light. The amount of the change depends on the intensity and duration of the exposure as well as the photosensitivity of the fibre. To write a high reflectivity fibre Bragg grating directly in the fibre the level of doping with germanium needs to be high. However, standard fibers can be used if the photosensitivity is enhanced by pre-soaking the fibre in hydrogen. More recently, fibre Bragg gratings have also been written in polymer fibers, this is described in the PHOSFOS entry.

Interference

This was the first method used widely for the fabrication of fibre Bragg gratings and uses two-beam interference. Here the UV laser is split into two beams which interfere with each other creating a periodic intensity distribution along the interference pattern. The refractive index of the photosensitive fibre changes according to the intensity of light that it is exposed to. This method allows for quick and easy changes to the Bragg wavelength, which is directly related to the interference period and a function of the incident angle of the laser light.

Sequential writing

Complex grating profiles can be manufactured by exposing a large number of small, partially overlapping gratings in sequence. Advanced properties such as phase shifts and varying modulation depth can be introduced by adjusting the corresponding properties of the subgratings. In the first version of the method, subgratings were formed by exposure with UV pulses, but this approach had several drawbacks, such as large energy fluctuations in the pulses and low average power. A sequential writing method with continuous UV radiation that overcomes these problems has been demonstrated and is now used commercially. The photosensitive fibre is translated by an interferometrically controlled airbearing borne carriage. The interfering UV beams are focussed onto the fibre, and as the fibre moves, the fringes move along the fibre by translating mirrors in an interferometer. As the mirrors have a limited range, they must be reset every period, and the fringes move in a sawtooth pattern. All grating parameters are accessible in the control software, and it is therefore possible to manufacture arbitrary gratings structures without any changes in the hardware.

Photomask

A photomask having the intended grating features may also be used in the manufacture of fibre Bragg gratings. The photomask is placed between the UV light source and the photosensitive fibre. The shadow of the photomask then determines the grating structure based on the transmitted intensity of light striking the fibre. Photomasks are specifically used in the manufacture of chirped Fibre Bragg gratings, which cannot be manufactured using an interference pattern.

Point-by-point

A single UV laser beam may also be used to 'write' the grating into the fibre point-by-point. Here, the laser has a narrow beam that is equal to the grating period. This method is specifically applicable to the fabrication of long period fibre gratings. Point-by-point is also used in the fabrication of tilted gratings.

Production

Originally, the manufacture of the photosensitive optical fibre and the 'writing' of the fibre Bragg grating were done separately. Today, production lines typically draw the fibre from the preform and 'write' the grating, all in a single stage. As well as reducing associated costs and time, this also enables the mass production of fibre Bragg gratings. Mass production is in particular facilitating applications in smart structures utilizing large numbers (3000) of embedded fibre Bragg gratings along a single length of fibre.

THEORY

The fundamental principle behind the operation of a FBG is Fresnel reflection,where light traveling between media of different refractive indices may both reflect and refract at the interface.

The refractive index will typically alternate over a defined length. The reflected wavelength (λB), called the Bragg wavelength, is defined by the relationship,

$$\lambda_B = 2ne\Lambda$$

where n_e is the effective refractive index of the grating in the fibre core and Λ is the grating period. The effective refractive index quantifies the velocity of propagating light as compared to its velocity in vacuum. n_e depends not only on the wavelength but also (for multimode waveguides) on the mode in which the light propagates. For this reason, it is also called modal index.

The wavelength spacing between the first minima, or the bandwidth (f****), is (in the strong grating limit) given by,

$$\Delta\lambda = \left[\frac{2\delta n_0 \eta}{\pi}\right]$$

where δn_0 is the variation in the refractive index $(n_3 - n_2)$, and η is the fraction of power in the core. Note that this approximation does not apply to weak gratings where the grating length, *Lg*, is not large compared to $\lambda_B \setminus \delta n_0$.

The peak reflection ($P_B(\lambda_B)$) is approximately given by,

$$P_B(\lambda_B) \approx \tanh^2\left[\frac{N\eta(V)\delta n_0}{n}\right]$$

where N is the number of periodic variations. The full equation for the reflected power ($P_B(\lambda)$), is given by,

$$P_B(\lambda) = \frac{\sinh^2\left[\eta(V)\delta n_0 \sqrt{1-\Gamma^2}\,\frac{N\Lambda}{\lambda}\right]}{\cosh^2\left[\eta(V)\delta n_0 \sqrt{1-\Gamma^2}\,\frac{N\Lambda}{\lambda}\right] - \Gamma^2}$$

where,

$$\Gamma(\lambda) = \frac{1}{\eta(V)\delta n_0}\left[\frac{\lambda}{\lambda_B} - 1\right]$$

TYPES OF GRATINGS

The term *type* in this context refers to the underlying photosensitivity mechanism by which grating fringes are produced in the fibre. The different methods of creating these fringes have a significant effect on physical attributes of the produced grating, particularly the temperature response and ability to withstand elevated temperatures. Thus far, five (or six) types of FBG have been reported with different underlying photosensitivity mechanisms. These are summarized below:

Standard, or type I, gratings

Written in both hydrogenated and non-hydrogenated fibre of all types, type I gratings are usually known as standard gratings and are manufactured in fibers of all types under all hydrogenation conditions. Typically, the reflection spectra of a type I grating is equal to 1-T where T is the transmission spectra. This means that the reflection and transmission spectra are complementary and there is negligible loss of light by reflection into the cladding or by absorption. Type I gratings are the most commonly used of all grating types, and the only types of grating available off-the-shelf at the time of writing.

Type IA gratings

- Regenerated grating written after erasure of a type I grating in hydrogenated germanosilicate fibre of all types

Type IA gratings were first observed in 2001 during experiments designed to determine the effects of hydrogen loading on the formation of IIA gratings in germanosilicate fibre. In contrast to the anticipated decrease (or 'blue shift') of the gratings' Bragg wavelength, a large increase (or 'red shift') was observed. Later work showed that the increase in Bragg wavelength began once an initial type I grating had reached peak reflectivity and begun to weaken. For this reason, it was labeled as a regenerated grating. Determination of the type IA gratings' temperature coefficient showed that it was lower than a standard grating written under similar conditions. The key difference between the inscriprion of type IA and IIA gratings is that IA gratings are written in hydrogenated fibres, whereas type IIA gratings are written in non-hydrogenated fibres.

Type IIA, or type In, gratings

- These are gratings that form as the negative part of the induced index change overtakes the positive part. It is usually associated with gradual relaxation of induced stress along the axis and/or at the interface. It has been proposed that these gratings could be relabeled type In (for type 1 gratings with a negative index change; type II label could be reserved for those that are distinctly made above the damage threshold of the glass).

Later research by Xie et al. showed the existence of another type of grating with similar thermal stability properties to the type II grating. This grating exhibited a negative change in the mean index of the fibre and was termed type IIA. The gratings were formed in germanosilicate fibers with pulses from a frequency doubled XeCl pumped dye laser. It was shown that initial exposure formed a standard (type I) grating within the fibre which underwent a small red shift before being erased. Further exposure showed that a grating reformed which underwent a steady blue shift whilst growing in strength.

Regenerated gratings

These are gratings that are reborn at higher temperatures after erasure of gratings, usually type I gratings and usually, though not always, in the presence of hydrogen. They have been interpreted in different ways including dopant diffusion (oxygen being the most popular current interpretation) and glass structural change. Recent work has shown that there exists a regeneration regime beyond diffusion where gratings can be made to operate at temperatures in excess of 1,295 °C, outperforming even type II femtosecond gratings. These are extremely attractive for ultra high temperature applications.

Type II gratings

- Damage written gratings inscribed by multiphoton excitation with higher intensity lasers that exceed the damage threshold of the glass. Lasers employed are usually pulsed in order to reach these intensities. They include recent developments in multiphoton excitation using femtosecond pulses where the short timescales (commensurate on a timescale similar to local relaxation times) offer unprecedented spatial localization of the induced change. The amorphous network of the glass is usually transformed via a different ionization and melting pathway to give either higher index changes or create, through micro-explosions, voids surrounded by more dense glass.

Archambault et al. showed that it was possible to inscribe gratings of ~100 per cent (>99.8 per cent) reflectance with a single UV pulse in fibers on the draw tower. The resulting gratings were shown to be stable at temperatures as high as 800 °C (up to 1,000 °C in some cases, and higher with femtosecond laser inscription). The gratings were inscribed using a single 40 mJ pulse from an excimer laser at 248 nm. It was further shown that a sharp threshold was evident at ~30 mJ; above this level the index modulation increased by more than two orders of magnitude, whereas below 30 mJ the index modulation grew linearly with pulse energy. For ease of identification, and in recognition of the distinct differences in thermal stability, they labeled gratings fabricated below the threshold as type I gratings and above the threshold as type II gratings. Microscopic examination of these gratings showed a periodic damage track at the grating's site within the fibre; hence type II gratings are also known as damage gratings. However, these cracks can be very localized so as to not play a major role in scattering loss if properly prepared

GRATING STRUCTURE

The structure of the FBG can vary via the refractive index, or the grating period. The grating period can be uniform or graded, and either localised or distributed in a superstructure. The refractive index has two primary characteristics, the refractive index profile, and the offset. Typically, the refractive index profile can be uniform or apodized, and the refractive index offset is positive or zero.

There are six common structures for FBGs;

1. Uniform positive-only index change,
2. Gaussian apodized,
3. Raised-cosine apodized,
4. Chirped,
5. Discrete phase shift, and
6. Superstructure.

The first complex grating was made by J. Canning in 1994. This supported the development of the first distributed feedback (DFB) fibre lasers, and also laid the groundwork for most complex gratings that followed, including the sampled gratings first made by Peter Hill and colleagues in Australia.

Apodized gratings

There are basically two quantities that control the properties of the FBG. These are the grating length, *Lg*, given as

$$L_g = N\Lambda$$

and the grating strength, $\delta n_0 \eta$. There are, however, three properties that need to be controlled in a FBG. These are the reflectivity, the bandwidth, and the side-lobe strength. As shown above, in the strong grating limit (*i.e.*, for large δn_0) the bandwidth depends on the grating strength, and not the grating length. This means the grating strength can be used to set the bandwidth. The grating length, effectively *N*, can then be used to set the peak reflectivity, which depends on both the grating strength and the grating length. The result of this is that the side-lobe strength cannot be controlled, and this simple optimisation results in significant side-lobes. A third quantity can be varied to help with side-lobe suppression. This is apodization of the refractive index change. The term apodization refers to the grading of the refractive index to approach zero at the end of the grating. Apodized gratings offer significant improvement in side-lobe suppression while maintaining reflectivity and a narrow bandwidth. The two functions typically used to apodize a FBG are Gaussian and raised-cosine.

Chirped fibre Bragg gratings

The refractive index profile of the grating may be modified to add other features, such as a linear variation in the grating period, called a chirp. The reflected wavelength changes with the grating period, broadening the reflected spectrum. A grating possessing a chirp has the property of adding dispersion—namely, different wavelengths reflected from the grating will be subject to different delays. This property has been used in the development of phased-array antenna systems and polarization mode dispersion compensation, as well.

Tilted fibre Bragg gratings

In standard FBGs, the grading or variation of the refractive index is along the length of the fibre (the optical axis), and is typically uniform across the width of the fibre. In a tilted FBG (TFBG), the variation of the refractive index is at an angle to the optical axis. The angle of tilt in a TFBG has an effect on the reflected wavelength, and bandwidth.

Long-period gratings

Typically the grating period is the same size as the Bragg wavelength, as shown above. For a grating that reflects at 1,500 nm, the grating period is 500 nm, using a refractive index of 1.5. Longer periods can be used to achieve much broader responses than are possible with a standard FBG. These gratings are called long-period fibre grating. They typically have grating periods on the order of 100 micrometers, to a millimeter, and are therefore much easier to manufacture.

APPLICATIONS

Communications

The primary application of fibre Bragg gratings is in optical communications systems. They are specifically used as notch filters. They are also used in optical multiplexers and demultiplexers with an optical circulator, or optical add-drop multiplexer (OADM). Figure 5 shows 4 channels, depicted as 4 colours, impinging onto a FBG via an optical circulator. The FBG is set to reflect one of the channels, here channel 4. The signal is reflected back to the circulator where it is directed down and dropped out of the system. Since the channel has been dropped, another signal on that channel can be added at the same point in the network.

A demultiplexer can be achieved by cascading multiple drop sections of the OADM, where each drop element uses an FBG set to the wavelength to be demultiplexed. Conversely, a multiplexer can be achieved by cascading multiple add sections of the OADM. FBG demultiplexers and OADMs can also be tunable. In a tunable demultiplexer or OADM, the Bragg wavelength of the FBG can be tuned by strain applied by a piezoelectric transducer. The sensitivity of a FBG to strain is discussed below in fibre Bragg grating sensors.

Fibre Bragg grating sensors

As well as being sensitive to strain, the Bragg wavelength is also sensitive to temperature. This means that fibre Bragg gratings can be used as sensing elements in optical fibre sensors. In a FBG sensor, the measurand causes a shift in the Bragg wavelength, $\Delta\lambda_B$. The relative shift in the Bragg wavelength,

$\Delta\lambda_B / \lambda_B$, due to an applied strain ($\in$) and a change in temperature (ΔT) is approximately given by,

$$\left[\frac{\Delta\lambda_B}{\lambda_B}\right] = C_S \in + C_T \Delta T$$

or,

$$\left[\frac{\Delta\lambda_B}{\lambda_B}\right] = (1 - pe) \in + (\alpha_\Lambda + \alpha_n)\Delta T$$

Here, C_S is the *coefficient of strain*, which is related to the *strain optic coefficient* f *pe*. Also, C_T is the *coefficient of temperature*, which is made up of the *thermal expansion coefficient* of the optical fibre, $\alpha\Lambda$, and the *thermo-optic coefficient*, αn.

Fibre Bragg gratings can then be used as direct sensing elements for strain and temperature. They can also be used as transduction elements, converting the output of another sensor, which generates a strain or temperature change from the measurand, for example fibre Bragg grating gas sensors use an absorbent coating, which in the presence of a gas expands generating a strain, which is measurable by the grating. Technically, the absorbent material is the sensing element, converting the amount of gas to a strain. The Bragg grating then transduces the strain to the change in wavelength.

Specifically, fibre Bragg gratings are finding uses in instrumentation applications such as seismology, pressure sensors for extremely harsh environments, and as downhole sensors in oil and gas wells for measurement of the effects of external pressure, temperature, seismic vibrations and inline flow measurement. As such they offer a significant advantage over traditional electronic gauges used for these applications in that they are less sensitive to vibration or heat and consequently are far more reliable. In the 1990s, investigations were conducted for measuring strain and temperature in composite materials for aircraft and helicopter structures.

Fibre bragg gratings used in fibre lasers

Recently the development of high power fibre lasers has generated a new set of applications for fibre Bragg gratings (FBG's), operating at power levels that were previously thought impossible. In the case of a simple fibre laser, the FBG's can be used as the high reflector (HR) and output coupler (OC) to form the laser cavity. The gain for the laser is provided by a length of rare earth doped optical fibre, with the most common form using Yb3+ ions as the active lasing ion in the silica fibre. These Yb-doped fibre lasers first operated at the 1 kW CW power level in 2004 based on free space cavities but were not shown to operate with fibre Bragg grating cavities until much later.

Such monolithic, all-fibre devices are produced by many companies worldwide and at power levels exceeding 1 kW. The major advantage of these all fibre systems, where the free space mirrors are replaced with a pair of fibre Bragg gratings (FBG's), is the elimination of realignment during the life of the system, since the FBG is spliced directly to the doped fibre and never needs adjusting. The challenge is to operate these monolithic cavities at the kW CW power level in large mode area (LMA) fibers such as 20/400 (20 um diameter core and 400 um diameter inner cladding) without premature failures at the intra-cavity splice points and the gratings. Once optimized, these monolithic cavities do not need realignment during the life of the device, removing any cleaning and degradation of fibre surface from the maintenance schedule of the laser. However, the packaging and optimization of the splices and FBGs themselves are non-trivial at these power levels as are the matching of the various fibers, since the composition of the Yb-doped fibre and various passive and photosensitive fibers needs to be carefully matched across the entire fibre laser chain. Although the power handling capability of the fibre itself far exceeds this level, and is possibly as high as >30 kW CW, the practical limit is much lower due to component reliability and splice losses.

Process of matching active and passive fibers

In a double-clad fibre there are two waveguides – the Yb-doped core that forms the signal waveguide and the inner cladding waveguide for the pump light. The inner cladding of the active fibre is often shaped to scramble the cladding modes and increase pump overlap with the doped core. The matching of active and passive fibers for improved signal integrity requires optimization of the core/clad concentricity, and the MFD through the core diameter and NA, which reduces splice loss. This is principally achieved by tightening all of the pertinent fibre specifications.

Matching fibers for improved pump coupling requires optimization of the clad diameter for both the passive and the active fibre. To maximize the amount of pump power coupled into the active fibre, the active fibre is designed with a slightly larger clad diameter than the passive fibers delivering the pump power. As an example, passive fibers with clad diameters of 395-um spliced to active octagon shaped fibre with clad diameters of 400-um improve the coupling of the pump power into the active fibre. An image of such a splice is shown, showing the shaped cladding of the doped double-clad fibre.

The matching of active and passive fibers can be optimized in several ways. The easiest method for matching the signal carrying light is to have identical NA and core diameters for each fibre. This however does not account for all the refractive index profile features. Matching of the MFD is also a method used to create matched signal carrying fibers. It has been shown that matching all of these components provides the best set of fibers to build high power amplifiers and lasers. Essentially, the MFD is modeled and the resulting target

NA and core diameter are developed. The core-rod is made and before being drawn into fibre its core diameter and NA are checked. Based on the refractive index measurements, the final core/clad ratio is determined and adjusted to the target MFD. This approach accounts for details of the refractive index profile which can be measured easily and with high accuracy on the preform, before it is drawn into fibre.

FIBRE CABLE TERMINATION

Fibre cable termination is the addition of connectors to each optical fibre in a cable. The fibers need to have connectors fitted before they can attach to other equipment. Two common solutions for fibre cable termination are pigtails and fanout kits or breakout kits.

PIGTAILS

A fibre pigtail is a single, short, usually tight-buffered, optical fibre that has an optical connector pre-installed on one end and a length of exposed fibre at the other end. The end of the pigtail is stripped and fusion spliced to a single fibre of a multi-fibre trunk. Splicing of pigtails to each fibre in the trunk "breaks out" the multi-fibre cable into its component fibers for connection to the end equipment. Pigtails can have female or male connectors. Female connectors could be mounted in a patch panel, often in pairs although single-fibre solutions exist, to allow them to be connected to endpoints or other fibre runs with patch fibers. Alternatively they can have male connectors and plug directly into an optical transceiver.

FANOUT KIT (BREAKOUT KIT)

A *fanout kit* is a set of empty jackets designed to protect fragile tight-buffered strands of fibre from a cable. This allows the individual fibers to be terminated without splicing, and without needing a protective enclosure such as a splicebox. This is normally an option with fibre distribution cable, or sometimes loose-buffer or ribbon cable, because these types of cable contain multiple strands that are designed for a permanent termination. Zip-cord style jackets, including those that contain Aramid yarn as the strength member, can be slipped over multiple fibre strands coming out of a loose buffer cable to convert it to a complete set of single-fibre cables that can be directly attached to optical connectors. A plastic boot is normally used for strain relief and protection from moisture. Use of a breakout kit enables a fibre-optic cable containing multiple loose buffer tubes to receive connectors without the splicing of pigtails.

FIBRE DISK LASER

A fibre disk laser is a fibre laser with transverse delivery of the pump light. They are characterized by the pump beam not being parallel to the active core of the optical fibre (as in a double-clad fibre), but directed to the coil of the fibre at an angle (usually, between 10 and 40 degrees). This allows use of the specific shape of the pump beam emitted by the laser diode, providing the efficient use of the pump. Fibre disk lasers should not be confused with the LaserDiscs (disk-shaped devices for storage and reading of information with laser beam) nor the disk laser or "active mirror", which is a laser with a thin active layer where the heat sink is realized in a direction opposite to that of propagation of the output beam.

REALIZATIONS OF FIBRE DISK LASERS

First disk lasers were developed in the Institute for Laser Science, Japan. Several realizations of fibre disk lasers were reported. The fibre disk laser is so named because the fibre is tightly coiled. Typically, no special feedback for the laser frequency is required, as the small reflection at end of the fibre is sufficient to provide efficient operation. In this case, both ends of the coiled fibre can be used as output.

APPLICATION AND POWER SCALING

Fibre disk lasers are used for cutting of metal (up to few mm thick), welding and folding. The disk-shaped configuration allows efficient heat dissipation (usually, the disks are cooled with flowing water)); allowing power scaling. When the increase of the length of the fibre becomes limited by stimulated scattering, additional power scaling can be achieved by combining several fibre disk lasers into a stack.

The spiral-coiled configuration is not the only possible arrangement; any other scheme of stacking of optical fibers with lateral delivery of pump can also be called a fibre disk laser, even if the resulting shape of the device is not circular. The term *fibre disk laser* applies to the concept of lateral delivery of pump to the active optical fibre rather than specifically to a disk-shaped device. The optimal shape of the fibre disk laser may depend on the properties of the beam of pump available, as well as on the specific application.

FIBRE TAPPING

Fibre tapping uses a network tap method that extracts signal from an optical fibre without breaking the connection. Tapping of optical fibre allows diverting some of the signal being transmitted in the core of the fibre into another fibre or a detector. Fibre to the home (FTTH) systems use beam splitters to allow many users to share one backbone fibre connecting to a central office, cutting the cost of each connection to the home. Test equipment can simply put a bend in the fibre and extract sufficient light to identify a fibre or determine if a signal is present.

Similar techniques can surreptitiously tap fibre for surveillance, although this is rarely done where electronic equipment used in telecommunication is required to allow access to any phone line for tapping by legal authorization. Tapping the fibre means that all signals from every communications source being routed through the fibre are presented and must be sorted for relevant data, an immense task when thousands of sources of data or voice may be present.

According to reports, tapping fibre was used by the US government for surveillance following the September 11, 2001 attacks and a nuclear submarine, the Jimmy Carter, was modified to allow tapping undersea cables.

DETECTING FIBRE TAPS

One way to detect fibre tapping is by noting increased attenuation added at the point of tapping. There are, however, tappers which allow tapping without significant added attenuation. In either case there should be a change of scattering pattern in that point in line which, potentially, can be detectable. However once the tapper has been detected it may be too late since a part of the information has been already eavesdropped.

COUNTER-MEASURES

One counter-measure is encryption to make the stolen data unintelligible to the thief. However, encryption can be an expensive solution, and there are also concerns about network bandwidth when it is used. Another counter-measure is to deploy a fibre-optic sensor into the existing raceway, conduit or armored cable. In this scenario, anyone attempting to physically access the data (copper or fibre infrastructure) is detected by the alarm system. A small number of alarm systems manufacturers provide a simple way to monitor the optical fibre for physical intrusion disturbances. There is also a proven solution that utilizes existing unused fibre (dark fibre) in a multi-strand cable for the purpose of creating an alarm system.

In the alarmed cable scenario, the sensing mechanism uses optical interferometry in which modally dispersive coherent light traveling through the multi-mode fibre mixes at the fiber's terminus, resulting in a characteristic pattern of light and dark splotches called a speckle pattern. The laser speckle is stable as long as the fibre remains immobile, but flickers when the fibre is vibrated. A fibre-optic sensor works by measuring the time dependence of this speckle pattern and applying digital signal processing to the Fast Fourier Transform (FFT) of the temporal data.

The U.S. government has been concerned about the tapping threat for many years, and it also has a concern about other forms of intentional or accidental physical intrusion. In the context of classified information Department of Defence (DOD) networks, Protective distribution system (PDS) is a set of military instructions and guidelines for network physical protection. PDS is defined a system of carriers (raceways, conduits, ducts, etc.) that are used to distribute Military and National Security Information (NSI) between two or more controlled areas or from a controlled area through an area of lesser classification, *i.e.*, outside the Sensitive Compartmented Information Facility (SCIF) or other similar area. National Security Telecommunications and Information Systems Security Instruction (NSTISSI 7003), Protective Distribution Systems (PDS), provides guidance for the protection of SIPRNet wire line and optical fibre PDS to transmit unencrypted classified National Security Information (NSI).

FILLING FACTOR

Filling factor, F, is a quantity measuring the efficiency of absorption of pump in the core of a double-clad fibre.

The efficiency of absorption of pumping energy in the fibre is an important parameter of a double-clad fibre laser. In many cases this efficiency can be approximated with

$$1-\exp\left(-F\frac{\pi r^2}{S}\alpha L\right)$$

where

S is the cross-sectional area of the cladding

r is the radius of the core (which is taken to be circular)

α is the absorption coefficient of pump light in the core

L is the length of the double-clad fibre, and

F is a dimensionless adjusting parameter, which is sometimes called the "filling factor"; $0 < F < 1$..

The filling factor may depend on the initial distribution of the pump light, the shape of the cladding, and the position of the core within it.

APPLICATION

The large (close to unity) filling factor is important in double-clad amplifiers; it allows them to reduce the requirements for the brightness of the pump and to reduce the length of the fibre laser. Such a reduction is especially important for the power scaling of various nonlinear processes, and contributions of stimulated scattering to the degradation of signal. Use of the filling factor for the estimate of the efficiency of absorption of the pump in fibre lasers allows quick estimates without performing complicated numerical simulations.

GRADED-INDEX FIBRE

In fibre optics, a graded-index or gradient-index fibre is an optical fibre whose core has a refractive index that decreases with increasing radial distance from the optical axis of the fibre. Because parts of the core closer to the fibre axis have a higher refractive index than the parts near the cladding, light rays follow sinusoidal paths down the fibre. The most common refractive index profile for a graded-index fibre is very nearly parabolic. The parabolic profile results in continual refocusing of the rays in the core, and minimizes modal dispersion. Multi-mode optical fibre can be built with either graded index or step index. The advantage of the graded index compared to step index is the considerable decrease in modal dispersion. This type of fibre is normalized by the International Telecommunications Union ITU-T at recommendation G.651.1.

PULSE DISPERSION

Pulse dispersion in a graded index optical fibre is given by

$$\text{Pulse dispersion} = \frac{k\delta n\, n_1\, l}{c},$$

where

δn is the difference in refractive indices of core and cladding,
n_1 is the refractive index of the cladding,
l is the length of the fibre taken for observing the pulse dispersion,
$c \approx 3 \times 10^8\, m/s$ is the speed of light, and
k is the constant of graded index profile.

HARD-CLAD SILICA OPTICAL FIBRE

Hard-clad silica (HCS) or polymer-clad fibre (PCF) is an optical fibre with a core of silica glass (diameter: 200 μm) and an optical cladding made of special plastic (diameter: 230 μm). In contrast to all-silica fibre, the core and cladding can be separated from each other.

Due to their medium bandwidths and transmission rates of less than 100 MBit/s, HCS fibers are suitable for distances of up to 2 km, *e.g.*. in local networks in buildings and industry. Generally, the following applies: The higher the attenuation, the shorter the distance.

Application area fiber types				
Fiber type	Core/Cladding	Application area	Distance	Data rate
Glass fiber	9/125 µm 10/125 µm	telecommunications	more than 10 km	MBit/s up to Gbit/s
Glass fiber	50/125 µm 62.5/125 µm	local networks in medium areas, buildings and telecommunications	up to 4 km	<155 MBit/s
HCS	200/230 µm	local networks in buildings and telecommunications	up to 2 km	<100 MBit/s
Plastic fiber (POF)	980/1000 µm	local networks in buildings, industry and automotive	up to 100 m	100 MBit/s

For comparison, plastic optical fibers (POF) have low bandwidths and transmission rates (typically 100 MBit/s). They also have a high attenuation and therefore, the maximum distance is around 100 meters. Glass fibers on the other hand have very high bandwidths and transmission rates of up to GBit/s. The attenuation in glass fibres is much lower, glass fibers can cover distances of more than 10 km. Regarding bandwidth and distances, HCS fibers are situated between POF and multimode or singlemode fibers.

INTRAMODAL DISPERSION

In fibre-optic communication, an intramodal dispersion, sometimes called material dispersion, is a category of dispersion that occurs within a single-mode. This dispersion mechanism is a result of material properties of optical fibre and applies to both single-mode and multi-mode fibers. There are two distinct types of intramodal dispersion: chromatic dispersion and polarization mode dispersion.

CHROMATIC DISPERSION

In silica, the index of refraction is dependent upon wavelength. Therefore different wavelengths will travel down an optical fibre at different velocities. This implies that a pulse with a wider FWHM will spread more than a pulse with a narrower FWHM. This dispersion limits both the bandwidth and the dis nce that information can be supported. This is why for long communications links, it is desirable to use a laser with a very narrow linewidth. Distributed Feedback (DFB) lasers are popular for communications because they have a single longitudinal mode with a very narrow

LONG-PERIOD FIBRE GRATING

A long-period fibre grating couples light from a guided mode into forward propagating cladding modes where it is lost due to absorption and scattering. The coupling from the guided mode to cladding modes is wavelength dependent so we can obtain a spectrally selective loss. It is an optical fibre structure with the properties periodically varying along the fibre, such that the conditions for the interaction of several copropagating modes are satisfied. The period of such a structure is of the order of a fraction of a millimeter. In contrast to the fibre Bragg gratings, LPFGs couple copropagating modes with close propagation constants; therefore, the period of such a grating can considerably exceed the wavelength of radiation propagating in the fibre. Because the period of an LPFG is much larger than the wavelength, LPFGs are relatively simple to manufacture. Since LPFGs couple copropagating modes, their resonances can only be observed in transmission spectra. The transmission spectrum has dips at the wavelengths corresponding to resonances with various cladding modes (in a single-mode fibre).

Depending on the symmetry of the perturbation that is used to write the LPFG, modes of different symmetries may be coupled. For instance, cylindrically symmetric gratings couple symmetric LP0m modes of the fibre. Microbend gratings, which are antisymmetric with respect to the fibre axis, create a resonance between the core mode and the asymmetric LP1m modes of the core and the cladding.

Long period grating has a wide variety of applications, including band-rejection filters, gain flattening filter and sensors.

Various gratings with complex structures have been designed: gratings combining several LPFGs, LPFGs with superstructures, chirped gratings, and gratings with apodization. Various LPFG-based devices have been developed: filters, sensors, fibre dispersion compensators, etc.

MICROSTRUCTURED OPTICAL FIBRE

Microstructured optical fibers (MOF) are optical fibre waveguides where guiding is obtained through manipulation of waveguide structure rather than its index of refraction.

In conventional optical fibers, light is guided through the effect of total internal reflection. The guiding occurs within a core of refractive index higher than refractive index of the surrounding material (cladding). The index change is obtained through different doping of the core and the cladding or through the use of different materials. In microstructured fibers, a completely different approach is applied. Fibre is built of one material (usually silica) and light guiding is obtained through the presence of air holes in the area surrounding the solid core. The holes are often arranged in the regular pattern in two dimensional arrays, however other patterns of holes exist, including non-periodic ones. While periodic arrangement of the holes would justify the use of term "photonic crystal fibre", the term is reserved for those fibers where propagation occurs within a photonic defect or due to photonic bandgap effect. As such, photonic crystal fibers may be considered a subgroup of microstructured optical fibers.

There are two main classes of MOF

1. Index guided fibers, where guiding is obtained through effect of total internal reflection
2. Photonic bandgap fibers, where guiding is obtained through constructive interference of scattered light (including photonic bandgap effect.)

Structured optical fibers, those based on channels running along their entire length go back to Kaiser and Co in 1974. These include air-clad optical fibers, microstructured optical fibers sometimes called photonic crystal fibre when the arrays of holes are periodic and look like a crystal, and many other subclasses.

Martelli and Canning realized that the crystal structures that have identical interstitial regions are actually not the most ideal structure for practical applications and pointed out aperiodic structured fibers, such as Fractal fibers, are a better option for low bend losses. Aperiodic fibers are a subclass of Fresnel fibers which describe optical propagation in analogous terms to diffraction free beams. These too can be made by using air channels appropriately positioned on the virtual zones of the optical fibre.

Photonic crystal fibers are a variant of the microstructured fibers reported by Kaiser et al. They are an attempt to incorporate the bandgap ideas of Yeh et al. in a simple way by stacking periodically a regular array of channels and drawing into fibre form. The first such fibers did not propagate by such a bandgap but rather by an effective step index – however, the name has, for historical reasons, remained unchanged although some researchers prefer to call these fibers "holey" fibers or "microstructured" optical fibers in reference to the pre-existing work from Bell Labs. The shift into the nanoscale was pre-empted by the more recent label "structured" fibers. An extremely important variant was the air-clad fibre invented by DiGiovanni at Bell Labs in 1986/87 based on work by Marcatili et al. in 1984. This is perhaps the single most successful fibre design to date based on structuring the fibre design using air holes and has important applications regarding high numerical aperture and light collection especially when implemented in laser form, but with great promise in areas as diverse as biophotonics and astrophotonics.

More interestingly, has been the recognition that the periodic structure is actually not the best solution for many applications. Fibers that go well beyond shaping the near field now can be deliberately designed to shape the far-field for the first time, including focusing light beyond the end of the fibre. These Fresnel fibers use well known Fresnel optics which has long been applied to lens design, including more advanced forms used in aperiodic, fractal, and irregular adaptive optics, or Fresnel/fractal zones. Many other practical design benefits include broader photonic bandgaps in diffraction based propagating waveguides and reduced bend losses, important for achieving structured optical fibers with propagation losses below that of step-index fibers.

MUXPONDER

In optical fibre communications, a muxponder is the element that sends and receives the optical signal on a fibre in much the same way as a transponder except that the muxponder has the additional functionality of multiplexing multiple sub-rate client interfaces onto the line interface.

NON-ZERO DISPERSION-SHIFTED FIBRE

Non-zero dispersion-shifted fibre (NZDSF), specified in ITU-T G.655, is a type of single-mode optical fibre which was designed to overcome the problems of dispersion-shifted fibre. NZDSF is available in two primary flavours: NZD+ and NZD-, which differ in their zero-dispersion wavelengths. These are typically around 1510 nm and 1580 nm, respectively. Because the zero-dispersion point of NZDSF is outside of the normal communications window, four-wave mixing and other non-linear effects are minimized. Other types of NZDSF include RS-NZDSF which has a reduced slope in its change of dispersion and large core NZDSF which further reduces residual non-linear distortion under high launch power. Some long-haul fibre paths will alternate NZD+ and NZD- segments to provide self-dispersion compensation with uniformly low dispersion across the minimum-loss window at 1550 nm.

OPTICAL FIBRE, NONCONDUCTIVE, RISER

Optical fibre, nonconductive, riser (OFNR) is a type of optical fibre cable. As designated by the National Fire Protection Association (NFPA), this name is used for interior fibre-optic cables which contain no electrically conductive components, and which are certified for use in riser applications; they are engineered to prevent the spread of fire from floor to floor in a building. Typically they are tested for compliance with ANSI/UL 1666-1997, *Standard Test for Flame Propagation Height of Electrical and Optical-Fibre Cable Installed Vertically in Shafts*. NFPA NEC 2005 Art 770.51(B) FPN. They are distinct from optical fibre, nonconductive, plenum cable (OFNP), and general-purpose optical cable. According to Colin Yao, "OFNR cables can not be installed in plenum areas since they do not have the required fire and smoking rating as Plenum rated cables. OFNP plenum cables can be used as substitutes for OFNR cables."

OPTICAL GROUND WIRE

An optical ground wire (also known as an OPGW or, in the IEEE standard, an optical fibre composite overhead ground wire) is a type of cable that is used in the construction of electric power transmission and distribution lines. Such cable combines the functions of grounding and communications. An OPGW cable contains a tubular structure with one or more optical fibers in it, surrounded by layers of steel and aluminum wire. The OPGW cable is run between the tops of high-voltage electricity pylons. The conductive part of the cable serves to bond adjacent towers to earth ground, and shields the high-voltage conductors from lightning strikes. The optical fibers within the cable can be used for high-speed transmission of data, either for the electrical utility's own purposes of protection and control of the transmission line, for the utility's own voice and data communication, or may be leased or sold to third parties to serve as a high-speed fibre interconnection between cities.

The optical fibre itself is an insulator and protects against power transmission line and lightning induction, external noise and cross-talk. Typically OPGW cables contain single-mode optical fibers with low transmission loss, allowing long distance transmission at high speeds. The outer appearance of OPGW is similar to ACSR cable usually used for shield wires.

An OPGW cable was patented by BICC in 1977 and installation of optical ground wires became widespread starting in the 1980s. In the peak year of 2000, around 60,000 km of OPGW was installed world-wide. Asia, especially China, has become the largest regional market for OPGW used in transmission-line construction.

CONSTRUCTION

Several different styles of OPGW are made. In one type, between 8 and 48 glass optical fibers are placed in a plastic tube. The tube is inserted into a stainless steel, aluminum, or aluminum-coated steel tube, with some slack length of fibre allowed to prevent strain on the glass fibers. The buffer tubes are filled with grease to protect the fibre unit from water and to protect the steel tube from corrosion, the interstices of the cable are filled with grease. The tube is stranded into the cable with aluminum, aluminium alloy or steel strands, similar to an ACSR cable. The steel strands provide strength, and the aluminum strands provide electrical conductivity. For very large fibre counts, up to 144 fibers in one cable, multiple tubes are used.

In other types, an aluminum rod has several spiral grooves around the outside, in which fibers in buffer tubes are laid. The fibre unit is covered with a plastic or steel tape, and the whole surrounded with aluminum and steel strands.

Individual fibers may be in "loose buffer" tubes, where the inside diameter of the tube is greater than the fibre outside diameter, or may be "tight buffered" where the plastic buffer is coated directly on to the glass. Fibers for OPGW are single-mode type.

COMPARISON WITH OTHER METHODS

Optical fibers are used by utilities as an alternative to private point to point microwave systems, power line carrier or communication circuits on metallic cables. OPGW as a communication medium has some advantages over buried optical fibre cable. Installation cost per kilometre is lower than a buried cable. Effectively, the optical circuits are protected from accidental contact by the high voltage cables below (and by the elevation of the OPGW from ground). A communications circuit carried by an overhead OPGW cable is unlikely to be damaged by excavation work, road repairs or installation of buried pipelines. Since the overall dimensions and weight of an OPGW is similar to the regular grounding wire, the towers supporting the line do not experience extra loading due to cable weight, wind and ice loads. An alternative to OPGW is use of the power cables to support a separately-installed fibre bundle. Other alternatives include fibre-bearing composite power conductors (OPCC), wrapped fibre optic cable (SkyWrap or OPAC), or using transmission towers to support a separate All-Dielectric Self-Supporting fibre cable with no conductive elements.

APPLICATION

A utility may install many more fibers than it needs for its internal communications both to allow for future needs and also to lease or sell to telecommunications companies. Rental fees for these "dark fibers" (spares) can provide a valuable source of revenue for the electrical utility. However, when rights-of-way for a transmission line have been expropriated from landowners, occasionally utilities have been restricted from such leasing agreements on the basis that the original right of way was only granted for electric power transmission.

Lower-voltage distribution lines may also carry OPGW wires for bonding and communications; however, utilities may also install all-dielectric self-supporting (ADSS) cables on distribution pole lines. These cables are somewhat similar to those used for telephone and cable television distribution.

While OPGW is easily installed in new construction, electrical utilities find the increased capacity of fibre to be so useful that techniques have been worked out for replacement of ground wires with OPGW on energized lines. Live-line working techniques are used to re-strand the towers with OPGW replacing the all-metal type of overhead shield wires.

INSTALLATION

Installation of OPGW requires some additional planning because it is impractical to splice an OPGW cable in mid-span; the lengths of cable purchased must be coordinated with the spans between towers to prevent waste. Where fibers must be joined between lengths, a weatherproof splice box is installed on a tower; a similar box is used to transition from the OPGW to an outside plant fibre-only cable to connect the fibers to terminal equipment.

PHOTONIC-CRYSTAL FIBRE

Photonic-crystal fibre (PCF) is a new class of optical fibre based on the properties of photonic crystals. Because of its ability to confine light in hollow cores or with confinement characteristics not possible in conventional optical fibre, PCF is now finding applications in fibre-optic communications, fibre lasers, nonlinear devices, high-power transmission, highly sensitive gas sensors, and other areas. More specific categories of PCF include photonic-bandgap fibre (PCFs that confine light by band gap effects), holey fibre (PCFs using air holes in their cross-sections), *hole-assisted* fibre (PCFs guiding light by a conventional higher-index core modified by the presence of air holes), and Bragg fibre (photonic-bandgap fibre formed by concentric rings of multilayer film). Photonic crystal fibers may be considered a subgroup of a more general class of microstructured optical fibers, where light is guided by structural modifications, and not only by refractive index differences.

Optical fibers have evolved into many forms since the practical breakthroughs that saw their wider introduction in the 1970s as conventional step index fibers and later as single material fibers where propagation was defined by an effective air cladding structure.

In general, regular structured fibers such as photonic crystal fibers, have a cross-section (normally uniform along the fibre length) microstructured from one, two or more materials, most commonly arranged periodically over much of the cross-section, usually as a "cladding" surrounding a core (or several cores) where light is confined. For example, the fibers first demonstrated by Russell consisted of a hexagonal lattice of air holes in a silica fibre, with a solid (1996) or hollow (1998) core at the center where light is guided. Other arrangements include concentric rings of two or more materials, first proposed as "Bragg fibers" by Yeh and Yariv (1978), a variant of which was recently fabricated by Temelkuran *et al.* (2002) and others.

(Note: PCFs and, in particular, Bragg fibers, should not be confused with fibre Bragg gratings, which consist of a periodic refractive index or structural variation along the fibre axis, as opposed to variations in the transverse directions as in PCF. Both PCFs and fibre Bragg gratings employ Bragg diffraction phenomena, albeit in different directions.)

The lowest reported attenuation of solid core photonic crystal fibre is 0.37 dB/km, and for hollow core is 1.2dB/km

CONSTRUCTION

Generally, such fibers are constructed by the same methods as other optical fibers: first, one constructs a "preform" on the scale of centimeters in size, and then heats the preform and draws it down to a much smaller diameter (often nearly as small as a human hair), shrinking the preform cross section but (usually) maintaining the same features. In this way, kilometers of fibre can be produced from a single preform. The most common method involves stacking, although drilling/milling was used to produce the first aperiodic designs. This formed the subsequent basis for producing the first soft glass and polymer structured fibers.

Most photonic crystal fibers have been fabricated in silica glass, but other glasses have also been used to obtain particular optical properties (such has high optical non-linearity). There is also a growing interest in making them from polymer, where a wide variety of structures have been explored, including graded index structures, ring structured fibers and hollow core fibers. These polymer fibers have been termed "MPOF", short for microstructured polymer optical fibers (van Eijkelenborg, 2001). A combination of a polymer and a chalcogenide glass was used by Temelkuran *et al.* (2002) for 10.6 μm wavelengths (where silica is not transparent).

MODES OF OPERATION

Photonic crystal fibers can be divided into two modes of operation, according to their mechanism for confinement. Those with a solid core, or a core with a higher average index than the microstructured cladding, can operate on the same index-guiding principle as conventional optical fibre — however, they can have a much higher effective- refractive index contrast between core and cladding, and therefore can have much stronger confinement for applications in nonlinear optical devices, polarization-maintaining fibers, (or they can also be made with much *lower* effective index contrast). Alternatively, one can create a "photonic bandgap" fibre, in which the light is confined by a photonic bandgap created by the microstructured cladding – such a bandgap, properly designed, can confine light in a *lower-index* core and even a hollow (air) core. Bandgap fibers with hollow cores can potentially circumvent limits imposed by available materials, for example to create fibers that guide light in wavelengths for which transparent materials are not available (because the light is primarily in the air, not in the solid materials). Another potential advantage of a hollow core is that one can dynamically introduce materials into the core, such as a gas that is to be analyzed for the presence of some substance. PCF can also be modified by coating the holes with sol-gels of similar or different index material to enhance its transmittance of light.

PLASTIC OPTICAL FIBRE

Plastic optical fibre (POF) (or Polymer optical fibre) is an optical fibre which is made out of plastic. Traditionally PMMA (acrylic) is the core material, and fluorinated polymers are the cladding material. Since the late 1990s however, much higher-performance POF based on perfluorinated polymers (mainly polyperfluorobutenylvinylether) has begun to appear in the marketplace.

In large-diameter fibers, 96 per cent of the cross section is the core that allows the transmission of light. Similar to traditional glass fibre, POF transmits light (or data) through the core of the fibre. The core size of POF is in some cases 100 times larger than glass fibre.

POF has been called the "consumer" optical fibre because the fibre and associated optical links, connectors, and installation are all inexpensive. Due to the attenuation and distortion characteristics of the traditional PMMA fibers are commonly used for low-speed, short-distance (up to 100 meters) applications in digital home appliances, home networks, industrial networks (PROFIBUS, PROFINET), and car networks (MOST). The perfluorinated polymer fibers are commonly used for much higher-speed applications such as data center wiring and building LAN wiring.

In relation to the future request of high-speed home networking, there has been an increasing interest in POF as a possible option for next-generation Gigabit/s links inside the house. To this end, several European Research projects are active, such as POF-ALL and POF-PLUS. Several standardization bodies at country, European and WW levels are currently developing Gigabit communication standards for POF aimed towards Home networking applications. It is expected the release at the beginning of 2012. One future Gigabit POF standard is based on multilevel PAM modulation a frame structure, Tomlinson-Harashima Precoding and Multilevel coset coding modulation. The combination of all these techniques has proven to be an efficient way of achieving low cost implementations at the same time that the transmission theoretical maximum capacity of the POF is approached. Since 2014 a full family of PHY transceivers are available in the market enabling the design and manufacturing of home networking equipment truly delivering Gigabit speeds into the home. Other alternatives are schemes like DMT, PAM-2 NRZ, DFE equalization or PAM-4. VDE standard was published in 2013. After the publication the IEEE ask VDE to withdrawn the specification and bring all the effort to IEEE. VDE withdrawn the specification and a CFI was presented to IEEE on March 2014. IEEE study group has been meeting sence then.

For telecommunications, the more difficult-to-use glass optical fibre is more common. This fibre has a core made of germania-doped silica. Although the actual cost of glass fibers are similar to the plastic fibre, their installed cost is much higher due to the special handling and installation techniques required.

One of the most exciting developments in polymer fibers has been the development of microstructured polymer optical fibers (mPOF), a type of photonic crystal fibre. POF fibre also has applications in sensing. It is possible to write fibre Bragg gratings in single and multimode POF. There are advantages in doing this over using silica fibre since the POF can be stretched further without breaking, some applications are described in the PHOSFOS project page.

POF in short:

1. PMMA and Polystyrene are used as fibre core, with refractive indices of 1.49 and 1.59 respectively.
2. Generally, fibre cladding is made of silicone resin (refractive index ~1.46).
3. High refractive index difference is maintained between core and cladding.
4. High numerical aperture.
5. Have high mechanical flexibility and low cost.
6. Attenuation loss is about 1 dB/m @ 650 nm.
7. Bandwidth is ~5 MHz-km @ 650 nm.

PLASTIC-CLAD SILICA FIBRE

In telecommunications and fibre optics, a plastic-clad silica fibre or polymer-clad silica fibre (PCS) is an optical fibre that has a silica-based core and a plastic cladding. The cladding of a PCS fibre should not be confused with the polymer overcoat of a conventional all-silica fibre. PCS fibers in general have significantly lower performance characteristics, particularly higher transmission losses and lower bandwidths, than all-glass fibers. The main applications of plastic-clad silica fibre are industrial, medical or sensing applications where cores that are larger than those used in standard data communications fibers are advantageous.

POLARIZATION CONTROLLER

A polarization controller is an optical device which allows one to modify the polarization state of light.

TYPES AND OPERATION

Polarization controllers can be operated without feedback, typically by manual adjustment or by electrical signals from a generator, or with automatic feedback. The latter allows for fast polarization tracking. A polarization controller can have the task of transforming a fixed, known polarization into an arbitrary one. Since polarization states are defined by two degrees of freedom, for example azimuth angle and ellipticity angle of the polarization state, such a polarization controller needs two degrees of freedom. The same holds for the task of transforming an arbitrary polarization into a fixed, known one.

More general is the transformation of an arbitrary polarization into another arbitrary polarization. This needs three degrees of freedom. Such a polarization controller can for example be obtained by placing on the optical path three rotatable waveplates in cascade: a first quarterwave plate, which is oriented to transform the incident elliptical polarization into linear polarization, a halfwave plate, which transforms this linear polarization into another linear polarization, and a second quarterwave plate, which transforms the other linear polarization into the desired elliptical output polarization.

Polarization controllers can be implemented with free space optics, through a fibre pigtailed U-bench, for example. In that case, light exits the fibre, passes through the three waveplates, that can be freely rotated to allow polarization adjustment and then enters back into the fibre. Polarization controllers can also be implemented in an all-fibre solution. In that case, the polarization of light is changed through the application of a controlled stress to the fibre itself.

For polarization controllers with automatic feedback, integrated optical lithium niobate ($LiNbO_3$) devices are very suitable. Polarization controllers with tracking speeds of up to 100 krad/s on the Poincaré sphere are commercially available (see external link at the bottom).

POLARIZATION-MAINTAINING OPTICAL FIBRE

In fibre optics, polarization-maintaining optical fibre (PMF or PM fibre) is a single-mode optical fibre in which linearly polarized light, if properly launched into the fibre, maintains a linear polarization during propagation, exiting the fibre in a specific linear polarization state; there is little or no cross-coupling of optical power between the two polarization modes. Such fibre is used in special applications where preserving polarization is essential.

POLARIZATION CROSSTALK

In an ordinary (non-polarization-maintaining) fibre, two polarization modes (say vertical and horizontal polarization) have the same nominal phase velocity due to the fiber's circular symmetry. However tiny amounts of random birefringence in such a fibre, or bending in the fibre, will cause a tiny amount of crosstalk from the vertical to the horizontal polarization mode. And since even a short portion of fibre, over which a tiny coupling coefficient may apply, is many thousands of wavelengths long, even that small coupling between the two polarization modes, applied coherently, can lead to a large power transfer to the horizontal mode, completely changing the wave's net state of polarization. Since that coupling coefficient was unintended and a result of arbitrary stress or bending applied to fibre, the output state of polarization will itself be random, and will vary as those stresses or bends vary; it will also vary with wavelength.

PRINCIPLE OF OPERATION

Polarization-maintaining fibers work by *intentionally* introducing a systematic linear birefringence in the fibre, so that there are two well defined polarization modes which propagate along the fibre with very distinct phase velocities. The beat length L_b of such a fibre (for a particular wavelength) is the distance (typically a few millimeters) over which the wave in one mode will experience an additional delay of one wavelength compared to the other polarization mode. Thus a length $L_b/2$ of such fibre is equivalent to a half-wave plate. Now consider that there might be a random coupling between the two polarization states over a significant length of such fibre. At point 0 along the fibre, the wave in polarization mode 1 induces an amplitude into mode 2 at some phase. However at point 1/2 L_b along the fibre, the same coupling coefficient between the polarization modes induces an amplitude into mode 2 which is now 180 degrees *out of phase* with the wave coupled at point zero, leading to cancellation. At point L_b along the fibre the coupling is again in the original phase, but at 3/2 L_b it is again out of phase and so on. The possibility of coherent addition of wave amplitudes through crosstalk over distances much larger than L_b is thus eliminated. Most of the wave's power remains in the original polarization mode, and exits the fibre in that mode's polarization as it is oriented at the fibre end. Optical fibre connectors used for PM fibers are specially keyed so that the two polarization modes are aligned and exit in a specific orientation.

Note that a polarization-maintaining fibre does not polarize light as a polarizer does. Rather, PM fibre maintains the linear polarization of linearly polarized light provided that it is launched into the fibre aligned with one of the fiber's polarization modes. Launching linearly polarized light into the fibre at a different angle will excite both polarization modes, conducting the same wave at a slightly different phase velocities. At most points along the fibre the net polarization will be an elliptically polarized state, with a return to the original polarization state after an integer number of beat lengths. Consequently, if visible laser light is launched into the fibre exciting both polarization modes, scattering of propagating light viewed from the side, is observed with a light and dark pattern periodic over each beat length, since scattering is preferentially perpendicular to the polarization direction.

DESIGNS

Several different designs are used to create birefringence in a fibre. The fibre may be geometrically asymmetric or have a refractive index profile which is asymmetric such as the design using an elliptical cladding as shown in the diagram. Alternatively, stress permanently induced in the fibre will produce stress birefringence; this may be accomplished using rods of another material included within the cladding. Several different shapes of rod are used, and the resulting fibre is sold under brand names such as "Panda" and "Bow-tie".

It is possible to create a circularly birefringent optical fibre just using an ordinary (circularly symmetric) single-mode fibre and twisting it, thus creating internal torsional stress. That causes the phase velocity of right and left hand circular polarizations to significantly differ. Thus the two circular polarizations propagate with little crosstalk in between them

APPLICATIONS

Polarization-maintaining optical fibers are used in special applications, such as in fibre optic sensing, interferometry and quantum key distribution. They are also commonly used in telecommunications for the connection between a source laser and a modulator, since the modulator requires polarized light as input. They are rarely used for long-distance transmission, because PM fibre is expensive and has higher attenuation than singlemode fibre. The output of a PM fibre is typically characterized by its polarization extinction ratio (PER)—the ratio of correctly to incorrectly polarized light, expressed in decibels. The quality of PM patchcords and pigtails can be characterized with a PER meter.

QUADRUPLY CLAD FIBRE

In fibre optics, a quadruply clad fibre is a single-mode optical fibre that has four claddings. Each cladding has a refractive index lower than that of the core. With respect to one another, their relative refractive indices are, in order of distance from the core: lowest, highest, lower, higher. A quadruply clad fibre has the advantage of very low macrobending losses. It also has two zero-dispersion points, and moderately low dispersion over a wider wavelength range than a singly clad fibre or a doubly clad fibre.

RADIATION EFFECTS ON OPTICAL FIBERS

When optical fibers are exposed to ionizing radiation such as energetic electrons, protons, neutrons, X-rays, ´-radiation, etc., they undergo 'damage'.

The term 'damage' primarily refers to the additional loss of the propagating optical signal leading to decreased power at the output end which could lead to premature failure of the component and or system. The loss of power or 'darkening' occurs because the chemical bonds forming the optical fibre core are disrupted by the impinging high energy resulting in the appearance of new electronic transition states giving rise to additional absorption in the wavelength regions of interest. Once radiation source is removed, the fibre returns to its original state to some extent (a process called recovery). The extent of damage is governed by the balance between defect generation (excess attenuation) on one hand and defect annihilation (recovery) on the other hand. If the dose rate is low, an equilibrium state (between attenuation and recovery) is reached with some degree of darkening. On the contrary if the dose rate is high, the utility of fibre depends on the overall induced attenuation and the recovery time. Understanding these radiation induced effects is important particularly for space based applications where optical fibers are being considered for use in increasing number of applications.

Intrinsic defects are present in the matrix of a single component glass material like pure silica. These include per-oxy linkages, POL (≡Si-O-O-Si≡) which are oxygen interstitials, and oxygen deficient centers, ODC (≡Si-Si≡) which are oxygen vacancies. When exposed to ionizing radiation, these sites trap holes to form per-oxy radicals, POR (≡Si-O-O.) and E' centers (≡Si.), respectively. In addition, rapidly cooled silica has strained ≡Si-O-Si≡ bonds, which are cleaved upon radiation to form non-bridging oxygen hole centers (NBOHC) depicted as ≡Si-O. and E' centers by trapping holes and electrons, respectively. When the glass contains a second network former with the same valence as silicon such as germanium, the difference in the electronegativities favours the dopant as a hole trap. Hence greater radiation damage occurs in doped silica glass. To improve radiation resistance of pure silica core fibers, it is necessary to minimize the number density of these intrinsic defects. Minimization of defects is achieved not only by reducing the incorporation of impurities in glass but also by controlling the input gas composition, optimizing the thermal history of glass at all stages of fibre manufacturing and optimizing the stress in the fibre core. Other strategies include incorporation of dopants (such as fluorine) in the core that minimize formation of defect centers discussed above.

All optical fibers undergo some darkening depending on a number of factors that include: ionization type, optical fibre core glass composition, operating wavelength, dose rate, total accumulated dose, temperature and power propagating through the core. Since attenuation is composition dependent, it is observed that fibers having pure silica cores and fluorine down doped claddings are amongst the most radiation hard fibers. The presence of dopants in the core such as germanium, phosphorus, boron, aluminum, erbium, ytterbium, thulium, holmium etc. compromises the radiation hardness of optical fibers. To minimize damage consequences, it is better to use a pure silica core fibre at higher operating wavelength, lower dose rate, lower total accumulated dose, higher temperature (accelerated recovery) and higher signal power (photo-bleaching). In addition to these intrinsic steps, external engineering may be required to shield the fibre from the effects of radiation.

Germanium-doped core fibers can be radiation hard even at high concentrations of germanium. Such fibers reach saturation, anneal well at higher temperatures and are also responsive to photo-bleaching. In case of phosphorus-doped core fibers, attenuation increases linearly with increasing phosphorus content and these fibers do not reach saturation. Recovery is very difficult even at higher temperatures. Boron, aluminum and all the rare-earth dopants significantly affect fibre loss. Radiation performances of various SM, MM and PM fibers manufactured by different vendors that were tested in wide range of radiation environments have been compiled.

RADIO OVER FIBRE

Radio over fibre (RoF) refers to a technology whereby light is modulated by a radio signal and transmitted over an optical fibre link to facilitate wireless access, such as 3G and WiFi simultaneous from the same antenna. In other words, radio signals are carried over fibre-optic cable. Thus, a single antenna can receive any and all radio signals (3G, Wifi, cell, etc..) carried over a single-fibre cable to a central location where equipment then converts the signals; this is opposed to the traditional way where each protocol type (3G, WiFi, cell) requires separate equipment at the location of the antenna.

Although radio transmission over fibre is used for multiple purposes, such as in cable television (CATV) networks and in satellite base stations, the term RoF is usually applied when this is done for wireless access.

In RoF systems, wireless signals are transported in optical form between a central station and a set of base stations before being radiated through the air. Each base station is adapted to communicate over a radio link with at least one user's mobile station located within the radio range of said base station. The advantage is that the equipment for WiFi, 3G and other protocols can be centralized in one place, with remote antennas attached via fibre optic serving all protocols. It greatly reduces the equipment and maintenance cost of the network.

RoF transmission systems are usually classified into two main categories (*RF-over-fibre* ; *IF-over-fibre*) depending on the frequency range of the radio signal to be transported.

(a) In RF-over-fibre architecture, a data-carrying RF (radio frequency) signal with a high frequency is imposed on a lightwave signal before being transported over the optical link. Therefore, wireless signals are optically distributed to base stations directly at high frequencies and converted from the optical to electrical domain at the base stations before being amplified and radiated by an antenna. As a result, no frequency up–down conversion is required at the various base stations, thereby resulting in simple and rather cost-effective implementation is enabled at the base stations.

(b) In IF-over-fibre architecture, an IF (intermediate frequency) radio signal with a lower frequency is used for modulating light before being transported over the optical link. Therefore, before radiation through the air, the signal must be up-converted to RF at the base station.

ADVANTAGES

- Low attenuation: It is a well known fact that signals transmitted on optical fibre attenuate much less than through other media, especially when compared to wireless medium. By using optical fibre, the signal will travel further, reducing the need of repeaters.
- Low complexity: RoF makes use of the concept of a remote station (RS). This station only consists of an optical-to-electrical (O/E) (and an optional frequency up or down converter), amplifiers, and the antenna. This means that the resource management and signal generation circuitry of the base station can be moved to a centralized location and shared between several remote stations, thus simplifying the architecture.
- Lower cost: Simpler structure of remote base station means lower cost of infrastructure, lower power consumption by devices and simpler maintenance all contributed to lowering the overall installation and maintenance cost. Further reduction can also be made by use of low-cost graded index polymer optical fibre (GIPOF)
- Future-proof: Fibre optics are designed to handle gigabits/second speeds which means they will be able to handle speeds offered by future generations of networks for years to come. RoF technology is also protocol and bit-rate transparent, hence, can be employed to use any current and future technologies. The most popular use for RF over fibre is for cable TV systems. It is impossible to run RF signals over copper cable to more than few hundred feet. They are transporting their entire CATV channel lineup over a single-fibre optic cable, because this way they can transport the signal for hundreds of km. It works like this: An electrical RF signal usually in the range of 54–870 MHz is converted to modulated light using RF 1310 nm or 1550 nm laser optics. The light travels over single-mode fibre to the fibre optic RF receiver where is converted back to electrical RF. Electrical RF is directly connected to a TV or set-top box. 1550 nm is more popular because it has less losses in the fibre and by using

fibre-optic amplifier known as EDFA it is possible to extend the transport distance. 1310 nm is losing about 0.35 db/KM of optical signal, 1550 nm is losing only 0.25 db/km. Optical budget between transmitter and receiver varies, depending on the transmitter power and receiver sensitivity.

APPLICATIONS

- Access to dead zones: An important application of RoF is its use to provide wireless coverage in the area where wireless backhaul link is not possible. These zones can be areas inside a structure such as a tunnel, areas behind buildings, Mountainous places or secluded areas such as jungles.
- FTTA (Fibre to the antenna): By using an optical connection directly to the antenna, the equipment vendor can gain several advantages like low line losses, immunity to lightning strikes/electric discharges and reduced complexity of base station by attaching lightweight optical-to-electrical (O/E) converter directly to antenna.

DEPLOYMENT

As of April 2012, AT and T had 3000 systems deployed in the USA in places like stadiums, shopping malls and inside buildings. "We continue to go very, very aggressively on distributing the antenna system solutions", said CEO Randall Stephenson in 2012. In China, systems are being widely deployed in industrial zones, harbours, hospitals and supermarkets. Plans are in place to expand into rural zones along rail lines, and in new residential and commercial construction spaces. It is believed China will be the leading user of the technology and this will bring down the cost of equipment.

REFRACTIVE INDEX PROFILE

A refractive index profile is the distribution of refractive indices of materials within an optical fibre. Some optical fibre has a step-index profile, in which the core has one uniformly-distributed index and the cladding has a lower uniformly-distributed index. Other optical fibre has a graded-index profile, in which the refractive index varies gradually as a function of radial distance from the fibre center. Graded-index profiles include power-law index profiles and parabolic index profiles.

REMOTE SKYLIGHTS

Remote Skylights are optical systems capable of providing natural light to unlit locations. Utilising an arrangement of parabolic reflectors and fibre optics, Remote Skylights are able to transport natural sunlight to areas that would otherwise be dark, or would require artificial illumination.

Remote Skylights are composed chiefly of a solar collection dish, a "heliotube" and a distribution dish. The collection and distribution dishes are both parabolic reflectors. The collection dish is connected to a mechanism which tracks the transit of the sun across the sky, so as to maximise the intensity of light falling upon it. The heliotube is a bundle of optical fibres that channel the collected sunlight from the collection dish to the distribution dish. Unlike a typical skylight, the heliotube means that the two dishes do not need to be adjacent to one another.

BENEFITS

Remote Skylights provide two key advantages over artificial illumination:

1. The transported light contains the frequencies necessary for photosynthesis. (Though it is reported that harmful UV rays are filtered out.)
2. No power is required to sustain the illumination. This means that (after construction) no harmful greenhouse gases are produced.

SPLICEBOX

A splice box (also known as splice distributor) is a housing in which fibre optic cables begin or end. Fibre optics are fanned out in splice boxes that are situated at the end of fibre optic transmission paths. The main components of a splice box are the splice cassette that picks up the fibers and their reserves, and the front panel which contains different connectors for transmitting signals via copper or fibre optic cables. The splice cassette is removable in order to assemble fibre optics with a splice unit. The front panel can also be removed to splice the fibers to various connectors. Since modern splice cassettes already contain a splice tray, a splice holder, couplings and pigtails, the installation of the cables is facilitated. So-called hybrid splice boxes do not only ensure data transmission via copper cables RJ45 or fibre optics, but they also ensure the power supply. That becomes especially important when a splice box needs to be installed in indoor or outdoor applications that are difficult to access.

STEP-INDEX PROFILE

For an optical fibre, a step-index profile is a refractive index profile characterized by a uniform refractive index within the core and a sharp decrease in refractive index at the core-cladding interface so that the cladding is of a lower refractive index. The step-index profile corresponds to a power-law index profile with the profile parameter approaching infinity. The step-index profile is used in most single-mode fibers and some multimode fibers.

A step-index fibre is characterized by the core and cladding refractive indices n_1 and n_2 and the core and cladding radii a and b. Examples of standard core and cladding diameters 2a/2b are 8/125, 50/125, 62.5/125, 85/125, or 100/

140 (units of μm). The fractional refractive-index change $\Delta = \frac{n_1 - n_2}{n_1} << 1$. The value of n_1 is typically between 1.44 and 1.46, and Δ is typically between 0.001 and 0.02.

Step-index optical fibre is generally made by doping high-purity fused silica glass (SiO_2) with different concentrations of materials like titanium, germanium, or boron.

Pulse dispersion in a step index optical fibre is given by

$$\text{pulse dispersion} \frac{\Delta n_1 \ell}{c}$$

where

Δ is the difference in refractive indices of core and cladding.
n_1 is the refractive index of core
ℓ is the length of the optical fibre under observation

$$c = 3 \times 10^8 \text{ ms}^{-1}$$

SUBWAVELENGTH-DIAMETER OPTICAL FIBRE

Subwavelength-diameter optical fibre (SDF or SDOF) is an optical fibre whose diameter is less than the wavelength of the light being propagated through the fibre. An SDOF usually consists of long thick parts (same as conventional optical fibres) at both ends, transition regions (tapers), where the fibre diameter gradually decreases down to the subwavelength value, and a subwavelength-diameter waist, which is the main acting part of an SDOF.

There is no general agreement on how these optical elements are to be called, different groups preferring to emphasize different properties of such fibres, sometimes even using different terms. The names in use include:

- Subwavelength waveguide, subwavelength optical wire, subwavelength-diameter silica wire, subwavelength diameter fibre taper
- (Photonic) wire waveguide, photonic wire, photonic nanowire, optical nanowires, optical fibre nanowires
- Tapered (optical) fibre, fibre taper
- Submicron-diameter silica fibre
- Ultrathin optical fibres
- Optical nanofibre
- Optical microfibres
- Submicron fibre waveguides
- Micro/Nano optical wires (MNOW)

The term *waveguide* can be applied not only to fibres, but also to other waveguiding structures such as silicon photonic subwavelength waveguides. The term *submicron* is often synonymic to *subwavelength* in this case, taking into account that the majority of experiments are carried out with the light with the wavelength between 0.8 and 1.6 μm; however for other wavelengths this may not be true. All the names including the prefix *nano-* are somewhat misleading, since it is usually applied to objects with dimensions on the scale of nanometers or tens of nanometers (cf. nanoparticle, nanotechnology). The characteristic behaviour of the SDOF—high intensity of the electromagnetic field both inside and outside the fibre, maximum confinement of light in transversal cross-section—appears when the fibre diameter is about half of the wavelength of light. That is why the term *subwavelength* is the most appropriate for these objects.

SDF FEATURES

High power in the evanescent field

The main peculiarity of an SDF is that in the waist region, a significant part of the light's power propagates outside the fibre. Rigorously, this follows from the application of Maxwell's equations to a waveguide with circular cross-section. In a simplified way, this may be explained by the following. Light is guided in waveguides by total internal reflection (TIR) occurring on the interface between the waveguide and surrounding media. During TIR, the light intensity does not fall down to zero immediately at the interface, but decreases exponentially (vanishes) in the adjacent medium (the light field outside the waveguide is called the evanescent field). The depth of penetration of light during TIR depends on the exact configuration, but it is usually greater than or on the order of the wavelength of light.

An SDF has a diameter which is smaller than or on the order of the wavelength of light. Since an SDF is also a waveguide and thus light propagation is physically explained by the same fundamental reasons as TIR, the light, guided by an SDF, penetrates into surrounding media (air or vacuum) to the depth of about one wavelength or more. However, while in the case of conventional waveguides this depth is very small compared to the waveguide dimensions and so only a negligible amount of energy propagates outside the waveguide, in the case of SDF the volume occupied by the evanescent field is bigger than the volume of the SDF itself. Therefore the evanescent field of an SDF contains a significant portion of the whole light energy propagating along the fibre.

MANUFACTURING

An SDF is usually created by tapering a commercial optical fibre. Special pulling machines accomplish the process. An optical fibre usually consists of a core, a cladding and a protective coating. Before pulling a fibre, its coating is removed (the fibre is stripped). Then the bare fibre is fixed at two ends on the movable translation stages of the pulling machine. The middle of the fibre between the stages is then heated with a flame or a laser beam and at the same time the translation stages move in the opposite directions. The glass melts and the fibre is elongated so that its diameter decreases. The flame or laser beam usually also moves in order to obtain waist of significant length and constant thickness. Using the described method, waists of 1...10 mm in length and diameters down to 100 nm are obtained.

HANDLING

Being extremely thin, an SDF is also extremely fragile. Therefore, an SDF is usually mounted onto a special frame immediately after pulling and is never detached from this frame. Another issue is dust particles which may adsorb to the surface of an SDF. If significant laser power is coupled into the fibre, dust particles will scatter light in the evanescent field, heat up and may thermally destroy the waist. In order to prevent this, SDF are pulled and used in dust-free environments such as flowboxes or vacuum chambers.

APPLICATIONS

- Sensors
- Nonlinear optics
- Fibre couplers
- Atom trapping and guiding.

Bibliography

A.F. Barker : *Handbook of Textiles*, Abhishek Publication, 2009.

Aashima Arora and Hemant Kapoor : *Home Textiles*, Bio-Green Books, 2012.

Ajeet Jaiswal : *Anthropo-Medical Profile of Textile Workers*, Alfa Pub, 2012.

Anita Tyagi : *Interior Textiles*, Sonali Publication, 2011.

Anne Morrell : *Indian Embroidery Techniques at the Calico Museum of Textiles*, Sarabhai Foundation, 2002.

Berenice Ellena : *India Sutra on the Magic Trail of Textiles*, Shubhi Publication, 2007.

Dick Kooiman : *Bombay Textile Labour: Managers, Trade Unionists and Officials 1918-1939*, Manohar, Publication, 2002

Fredrick W Bunce : *Buddhist Textiles of Laos, Lan Na and the Isan : The Iconography of Design Elements*, D K Printworld, 2003.

G.K. Ghosh and Shukla Ghosh : *Ikat Textiles of India*, APH, Publication, 2014.

G.K. Ghosh and Shukla Ghosh : *Indian Textiles : Past and Present*, APH, Publication, 2011.

J. M. Matthews : *Applications of Dyestuffs : To Textiles, Paper, Leather and Other Materials*, Abhishek Publication, 2014.

Jay Narayan Vyas : *Indian Textiles 2015 : Comprehensive Forecast on Indian Textiles Industry in 2015*, Textile Review, 2011.

John Gillow and Nicholas Barnard : *Indian Textiles*, Om Books International, 2008.

John Irwin : *Journal of Indian Textile History*, Calico Museum, 2002.

Martand Singh : *Handcrafted Indian Textiles*, Roli, Publication, 2005.

P Sarah : *Lasers and Optical Fibre Communications*, I.K. International, 2009.

P. Prince Dhanaraj : *Industrial Economics (With Special Reference to Cotton Textile Mills Industry)*, Reliance, 2002.

P.V. Vidyasagar : *Handbook of Textiles*, Mittal, Publication, 2005.

R.M. Lehri : *Indian Textiles* Super Book House, 2009.

Rajesh Kalra : *Bleaching of Textile Fibres*, Abhishek Publications, 2011.

Robin Mathew : *Textile and Apparel Industry*, Book Enclave, 2012.

Robin Mathew : *Textile and Apparel Industry*, Book Enclave, 2012.

Rosemary Crill : *Chintz Indian Textiles for the West*, Mapin, Publication, 2008.

Satya Prakash Sangar : *Indian Textiles in the Seventeenth Century*, Reliance, 2004.

Shuji Uchikawa : *Indian Textile Industry : State Policy, Liberalization and Growth,* Manohar, Publication, 2002.

Swati Bhargava : *Clothing and Textiles : (A Practical Aspect)*, Shree Niwas Publications, 2010.

William H. Dooley : *A Compendium of Textiles*, Abhishek Publication, 2011.

Index